WORKSHEETS
FOR CLASSROOM OR LAB PRACTICE
EDUTORIAL SERVICES, LLC

PREALGEBRA
FOURTH EDITION

Jamie Blair
Orange Coast College

John Tobey
North Shore Community College

Jeffrey Slater
North Shore Community College

Prentice Hall
is an imprint of

PEARSON

Copyright ©2010 Pearson Education, Inc.
Publishing as Pearson Prentice Hall, Upper Saddle River, NJ 07458.

ISBN-13: 978-0-321-58839-5
ISBN-10: 0-321-58839-8

2 3 4 5 6 BB 11

Prentice Hall
is an imprint of

www.pearsonhighered.com

Worksheets
For Classroom or Lab Practice

Prealgebra, Fourth Edition

Table of Contents

Practice Set 1.1
Understanding Whole Numbers

Write a word name for each number. Then write the number in expanded notation.

1. 5321

1. _____

_____

_____

2. 80,059

2. _____

_____

_____

3. 413,204

3. _____

_____

_____

Show how to request each of the following dollar amounts, using the minimum number of one-, ten-, and hundred-dollar bills.

4. $259

4. _____

5. $807

5. _____

Replace each question mark with an inequality symbols < or >.

6. 65 ? 64

6. _____

7. 1345 ? 1354

7. _____

8. 11,032 ? 10,032

8. _____

Rewrite using numbers and an inequality symbol.

9.   One million is greater than zero.   9.   _____

10. Ninety-eight is less than one hundred fourteen.   10. _____

11. Two thousand twelve is greater than one thousand nine hundred   11. _____
     six.

Round each of the following numbers to the place indicated.

12. 75 to the nearest ten   12. _____

13. 463 to the nearest ten   13. _____

14. 781 to the nearest hundred   14. _____

15. 2940 to the nearest hundred   15. _____

16. 879,931 to the nearest thousand   16. _____

Name _____   Date _____

Adding Whole Number Expressions

Use the indicated property to rewrite each sum, then simplify if possible.

1. $3 + y$; commutative property of addition

1. _____

2. $(a + 2) + 7$; associative property of addition

2. _____

3. $5 + (6 + x)$; associative property of addition

3. _____

4. $2781 + 385$; commutative property of addition

4. _____

Evaluate each expression using the given values of x and y.

5. $7 + x + y$ when $x = 2$ and $y = 4$

5. _____

6. $x + y + 18$ when $x = 13$ and $y = 9$

6. _____

7. $x + 64 + y$ when $x = 33$ and $y = 3$

7. _____

8. $y + 200 + x$ when $x = 645$ and $y = 1000$

8. _____

9. $x + y + 3400 + x$ when $x = 2000$ and $y = 1749$

9. _____

Name _____ Date _____

Add.

10.　　1 6 7
　　　+　4 1

10. _____

11.　　8 3 4
　　　+7 8 8

11. _____

12.　　3 7 5 4
　　　　3 5 8
　　　+6 8 3 9

12. _____

13.　　5 0 0 0
　　　　4 0 3
　　　　4 9 7 6
　　　+8 2 3 5

13. _____

Find the perimeter of each figure.

14.

15 in.

4 in.

14. _____

15.

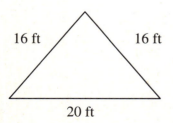

16 ft　　16 ft

20 ft

15. _____

4

Practice Set 1.3
Subtracting Whole Number Expressions

Translate each of the following into numbers and symbols. Then evaluate the expression.

1. nine minus eight   1. _____

2. the difference of forty-seven and twelve   2. _____

3. thirty-six less than one hundred twenty-three   3. _____

4. three hundred one subtracted from four hundred   4. _____

Evaluate each expression using the given values of x and y.

5. $11 - x - y$ when $x = 3$ and $y = 6$ 5. _____

6. $38 - y - x$ when $x = 10$ and $y = 15$ 6. _____

7. $x - 72 - y$ when $x = 135$ and $y = 29$ 7. _____

8. $x - y - 99$ when $x = 700$ and $y = 101$ 8. _____

Subtract and check.

9. $\begin{array}{r} 4\ 3\ 8\ 9 \\ -\ \ 6\ 9\ 8 \\ \hline \end{array}$ 9. _____

10. $\begin{array}{r} 3\,1,0\,0\,2 \\ -\,1\,4,2\,9\,8 \end{array}$ 10. _____

11. $\begin{array}{r} 1\,0\,9,0\,0\,0 \\ -\;\;6\,7,8\,9\,8 \end{array}$ 11. _____

Use subtraction to solve the following applied problems.

12.   Alicia earned $437 her first week working at Best Electric   12. _____
      Company. Stacey, though, earned only $198 her first week.
      How much more money did Alicia earn than Stacey?

13.   Leonard spent $755 on a washing machine, but received $125   13. _____
      cash back in the form of a rebate. How much did the washing
      machine cost him with the rebate?

14.   Cassandra's highest bowling score is 179 and Maura's is 206. By   14. _____
      how many points is Maura's best score higher than Cassandra's?

15.   I live 1720 miles from Lake Okeechobee but 2405 miles from   15. _____
      Lake Tahoe. How many miles farther from my house is Lake
      Tahoe than Lake Okeechobee?

16.   Barbara Ann received $175 in birthday money from her relatives.   16. _____
      She spent $35 on a new hairstyle, $26 to fix the front tire on her
      bicycle, and $59 on new clothes. How much money was left after
      all of these purchases?

Name _____   Date _____

Practice Set 1.4
Multiplying Whole Number Expressions

State what property of multiplication is represented in each mathematical statement.

1. $3 \cdot (4 \cdot 2) = (3 \cdot 4) \cdot 2$ 1. _____

2. $2 \cdot x = x \cdot 2$ 2. _____

Translate using numbers and symbols. Do not evaluate.

3. seven times eight   3. _____

4. the product of five and fifteen   4. _____

5. four doubled   5. _____

6. four times a number   6. _____

7. the product of a number and three   7. _____

Multiply.

8. 8(235)   8. _____

9. 35(400)   9. _____

10.   2 3 0 7
     ×   4 8 7

10. _____

11.   1 2, 4 0 2
     ×   6 0 7

11. _____

12. Rita bought three boxes of blank CDs. If each box holds 25 CDs,   12. _____
     how many CDs did Rita buy?

13. If a rectangular living room measures 13 feet in one direction and   13. _____
     17 feet in the other, how many square feet of carpeting will be
     needed to cover the entire floor?

14. Anna has five children. Each of her children has three children.   14. _____
     Also, each of her grandchildren has four children. How many
     great-grandchildren does Anna have?

15. The most recent shipment of hardware to Johnson Construction   15. _____
     Company contained three flats of nails. Each flat contained 12
     cases, each case contained 20 boxes, and each box contained 200
     nails. How many nails were in the shipment?

Name _____   Date _____

Divide.

1. $18 \div 9$ 1. _____

2. $163 \div 1$ 2. _____

3. $77 \div 0$ 3. _____

4. $6\overline{)102}$ 4. _____

5. $3\overline{)413}$ 5. _____

6. $8\overline{)1325}$ 6. _____

7. $21\overline{)2063}$ 7. _____

8.   41)8671   8. _____

Translate using numbers and symbols.

9.   thirty-six cookies divided equally among nine people   9. _____

10.   the quotient of seventy-two and eight   10. _____

11.   fifty-two divided by a number   11. _____

Solve the following word problems.

12.   Keith planted eight rows of tomatoes all containing the same   12. _____
      number of plants. He planted 96 tomato plants all together. How
      many plants were in each row?

13.   Amy, Alicia, Amber, and Alison pooled their allowances and   13. _____
      purchased a box of 46 candy bars. They agreed to split the box
      evenly among the four of them and give any leftover candy to
      their friend Stacey. How many candy bars did Stacey receive?

14.   Corey worked three different jobs last month and accumulated a   14. _____
      total of 168 hours. If he worked the same number of hours at
      each job, how many hours did he spend at each job?

15.   Jocasta has a rigorous exercise routine that lasts exactly 108   15. _____
      minutes every morning. In this time, she does nine different
      exercises, each for the same amount of time. How many minutes
      does she spend on each exercise?

16.   In preparation for a talent show, the drama club at Wesley High   16. _____
      School set up 396 folding chairs in the gym. They arranged the
      chairs in 18 equal rows. How many chairs were in each row?

Practice Set 1.6
Exponents and the Order of Operations

Write each product in exponent form. Do not evaluate.

1. $3 \times 3 \times 3 \times 3 \times 3$ 1. _____

2. $6 \times 6 \times 6$ 2. _____

3. $10 \times 10 \times 10 \times 10$ 3. _____

4. $a \times a \times a \times a \times a$ 4. _____

Translate using numbers and exponents. Do not evaluate.

5. nine squared   5. _____

6. eleven cubed   6. _____

7. y to the sixth power   7. _____

Evaluate.

8. 6^2 8. _____

9. 8^3 9. _____

10. 10^7 10. _____

11. $2^3 + 4^3 + 8^2$ 11. _____

12. $6^2 + 1^5 + 13^1$ 12. _____

Using the correct order of operations, evaluate each expression.

13. $3 \times 8 + 4 - 3$ 13. _____

14. $4 \times 10 - 20 \div 5 + 7^2$ 14. _____

15. $5 \times 2^3 - 4 \times (14 \div 7)$ 15. _____

16. $\dfrac{54(13-10)-18}{3^2-3}$ 16. _____

Name _____   Date _____

Translate using numbers and symbols. Do not evaluate or simplify.

1.   three times x plus 6   1. _____

2.   ten times the sum of five and eight   2. _____

3.   eight times the difference of x and one   3. _____

4.   the sum of y and five times three   4. _____

Evaluate for the given values.

5. $3x + 2y$ when $x = 4$ and $y = 5$ 5. _____

6. $7a - 9b$ when $a = 8$ and $b = 2$ 6. _____

7. $\dfrac{(a+8)}{2}$ when $a = 12$ 7. _____

8. $\dfrac{(m^2 - 5)}{4}$ when $m = 5$ 8. _____

9. $\dfrac{(x^2 + 6)}{y}$ when $x = 7$ and $y = 5$ 9. _____

10. $3mn + 2m + 5n$ when $m = 4$ and $n = 8$ 10. _____

Use the distributive property to simplify.

11. $5(a + 1)$ 11. _____

12. $9(y + 3)$ 12. _____

13. $8(a - 1 - 2b)$ 13. _____

14. $3(x - 5) + 21$ 14. _____

15. $4(x + y + 2) + y$ 15. _____

16. $3(x + y + 4) - 2(y + 1)$ 16. _____

Practice Set 1.8
Introduction to Solving Linear Equations

Combine like terms.

1. $5a + 12a$ 1. _____

2. $12m - 2m - 3m$ 2. _____

3. $3xy + 8xy + 2xy$ 3. _____

4. $7a + 2b + 3a + 9b + a$ 4. _____

5. $14ab + 8 + 4ab + 3$ 5. _____

Translate the mathematical symbols using words.

6. $6x$ 6. _____

    _____

7. $4a = 24$ 7. _____

    _____

8. $x + 8 = 12$ 8. _____

    _____

Name _____   Date _____

Solve and check your answer.

9. $x + 4 = 10$ 9. _____

10. $10 - n = 2$ 10. _____

11. $2x + x = 9$ 11. _____

12. $(a + 5) + 2 = 16$ 12. _____

Translate to an equation. Do not solve the equation.

13. Three plus what number equals twelve?   13. _____

14. If a number is subtracted from fourteen, the result is eleven.   14. _____

15. What number plus the sum of five and eight equals sixteen?   15. _____

16. If five is multiplied by the difference of ten and a number, the   16. _____
     result is thirty.

Practice Set 1.9
Solving Applied Problems Using Several Operations

Karen wants to buy the following four items:

ITEM	PRICE
Bedspread	$189
Digital camera	$331
Fountain	$94
Stained glass	$453

Use this chart to solve the following problems.

1. Round each item to the nearest hundred dollars.   1. _____

2. Using these rounded numbers, estimate the total cost of all four   2. _____
   items.

3. Using these rounded numbers, estimate how much more the   3. _____
   stained glass costs than the fountain.

Recently, the Catherine College Book Company reported the following textbook sales for
their top three sales associates during the fall and spring semesters:

NAME	FALL SEMESTER	SPRING SEMESTER
Chris	632	243
Dave	215	305
Ray	481	213

Use this chart to solve the following problems.

4. How many textbooks were sold during the fall semester?   4. _____

5. How many more books did Chris sell than Ray during both   5. _____
   semesters combined?

6. Who sold the fewest books for both semesters combined?   6. _____

Name _____   Date _____

Doris received an invoice from a hardware store:

NUMBER	ITEM	UNIT PRICE	TOTAL PRICE
1	rotary saw	$267.00	$267.00
3	hammer	$10.25	$30.75
7	paintbrush	$1.79	
9	sandpaper		$9.72
	paint roller	$2.59	$15.54

Unfortunately, some of the numbers have been smudged out. Use this chart to solve the following problems.

7. How much did Doris pay all together for her paintbrushes?   7. _____

8. How much does one piece of sandpaper cost?   8. _____

9. How many paint rollers did Doris buy?   9. _____

10. What was the total cost of all items purchased?   10. _____

Solve the following problems.

11. Last week, Maureen worked 53 hours. She earns $10 per hour for the first 40 hours worked, and then earns $15 per hour every hour after 40 hours worked. How much money should Maureen expect to be paid for last week's work?   11. _____

12. A textbook author estimates that she can write 3 pages per day. How many pages can she expect to write in 30 days? 60 days?   12. _____

13. An apple orchard has 23 rows of 15 trees. Each tree can expect to produce 350 apples. How many apples can the orchard owner expect to produce?   13. _____

14. Charles withdrew $200 from the bank to do some shopping. He spent $68.73 at the supermarket, then bought three folding tables for $21.69 each. Finally, he bought five boxes of Girl Scout cookies, paying $4.75 for each box. How much money did Charles have left after all of these purchases?   14. _____

Practice Set 2.1
Understanding Integers

Graph the following values on a number line.

1.   −5, −2, 0 3, 4

1.
$$\xleftarrow{\;\;\;}\underset{\substack{-5\;-4\;-3\;-2\;-1\;\;0\;\;1\;\;2\;\;3\;\;4\;\;5}}{\overline{|\;|\;|\;|\;|\;|\;|\;|\;|\;|\;|}}\xrightarrow{\;\;\;}$$

2.   −4, −2, 2, 4

2. $$\xleftarrow{\;\;\;}\underset{\substack{-5\;-4\;-3\;-2\;-1\;\;0\;\;1\;\;2\;\;3\;\;4\;\;5}}{\overline{|\;|\;|\;|\;|\;|\;|\;|\;|\;|\;|}}\xrightarrow{\;\;\;}$$

3.   −3, −1, 0, 1, 3

3. $$\xleftarrow{\;\;\;}\underset{\substack{-5\;-4\;-3\;-2\;-1\;\;0\;\;1\;\;2\;\;3\;\;4\;\;5}}{\overline{|\;|\;|\;|\;|\;|\;|\;|\;|\;|\;|}}\xrightarrow{\;\;\;}$$

State both the opposite and the absolute value.

4.   −21

4.   _____

5. $|{-}100|$

5.   _____

6. $-(-(-x))$ Assume x is a positive number.

6.   _____

Replace each ___ with the symbol <, >, or =.

7.   6 ___ −6

7.   _____

8. $-(-(14))$ ___ −14

8.   _____

9. $|{-}5|$ ___ −3

9.   _____

10. $-(-4) + |-3|$ ___ $|-5| + |-2|$ 10. _____

Evaluate using the given value of *x*.

11. Evaluate $x + (-x) - |x|$ for $x = 5$ 11. _____

12. Evaluate $|x| - 2 - |x - 2|$ for $x = -1$.   12. _____

13. Evaluate $-(-(-x)) - (-|x|)$ for $x = -21$.   13. _____

Answer each question with a short explanation.

14. Which number has the greater absolute value: 14. _____
     37 or –37? Explain your thinking.
                                                         _____

15. Which number has the greater opposite: –4 or   15. _____
     5? Explain your thinking.
                                                         _____

16. Which number is greater: the opposite of 1 or   16. _____
     the absolute value of 1? Explain your thinking.
                                                         _____

Practice Set 2.2
Adding Integers

Add using the rules for addition of integers.

1. $(-5) + (-3)$ 1. _____

2. $(-7) + 0$ 2. _____

3. $6 + (-11)$ 3. _____

4. $63 + (-21)$ 4. _____

5. $-20 + 20$ 5. _____

6. $35 + (-10) + (-16) + 8$ 6. _____

7. $(-15) + (-20) + (-6) + 40$ 7. _____

Evaluate these algebraic expressions.

8.   Evaluate $x + (-12)$ for $x = 25$.   8. _____

9.   Evaluate $-17 + (-x)$ for $x = -10$.   9. _____

10. Evaluate $-20 + (-x) + (-1)$ for $x = -15$.   10. _____

11. Evaluate $40 + x + (-y)$ for $x = 10$ and $y = -20$.   11. _____

12. Evaluate $-x + (-y) + (-100)$ for $x = 3$ and $y = -5$.   12. _____

Solve the following problems.

13. At 9 A.M. the temperature was 20°. By 3 P.M. the temperature had   13. _____
    dropped 10°. By 9 P.M. the temperature had dropped another 12°.
    Represent the temperature at 9 P.M. as an integer.

14. Jane started the year with $5000 in savings. During the first   14. _____
    quarter, she experienced a loss of $3000. During the second
    quarter, she lost an additional $4000. Represent her savings at
    the end of the second quarter as an integer.

15. On Friday morning, Kevin began his camping trip at an altitude   15. _____
    of 400 feet. He descended 350 feet on Friday, descended another
    75 feet on Saturday, and then climbed up 200 feet on Sunday.
    Represent his altitude on Sunday night as an integer.

Practice Set 2.3
Subtracting Integers

Subtract.

1.   15 − 20

2.   −10 − 36

3.   13 − (−10)

4.   −12 − (−20)

5.   −6 − 2 − 8

6.   −7 + (−11) − (−15)

Evaluate these algebraic expressions.

7.   Evaluate 5 − x for $x = 13$.

8.   Evaluate −6 − x − 3 for $x = -10$.

9.   Evaluate −x − 14 − 2 for $x = 5$.

1. _____

2. _____

3. _____

4. _____

5. _____

6. _____

7. _____

8. _____

9. _____

Name _____   Date _____

10. Evaluate $-x - (-11) - y$ for $x = 10$ and $y = -10$.   10. _____

11. Evaluate $-9 - x + y$ for $x = -8$ and $y = -6$.   11. _____

12. Evaluate $-5 - (-x) + (-4) - y - 7$ for $x = 1$ and $y = -1$.   12. _____

Solve the following problems.

13. On New Year's Day, the high temperature in Honolulu, Hawaii   13. _____
     was 84° and the low temperature in Helena, Montana was −14°.
     What is the difference between these two temperatures?

14. Ariel lives in a Manhattan penthouse that's 265 feet above street   14. _____
     level. Pete lives in the same building in a basement apartment
     that's 20 feet below street level. What is the difference in altitude
     between Ariel's apartment and Pete's?

15. Martha left New York by plane at 2:00 P.M. Her flight to San   15. _____
     Francisco took six hours, but the local time in San Francisco is
     three hours earlier than in New York. What was the local time in
     San Francisco when Martha's plane landed?

16. Irene has $742 in her checking account, but Justin is overdrawn   16. _____
     and owes the bank $47. How much more money does Irene have
     than Justin?

24

Name _____   Date _____

Practice Set 2.4
Multiplying and Dividing Integers

Find the product by writing as repeated addition.

1. 2(–10)   1. _____

2. 4(–7)   2. _____

3. 6(–1)   3. _____

Perform each operation indicated.

4. –4(8)   4. _____

5. (–5)(–9)   5. _____

6. $(-3)^2$ 6. _____

7. $(-5)^3$ 7. _____

8. 45 ÷ (–9)   8. _____

9. –100 ÷ 100   9. _____

10. (–64) ÷ (–8)   10. _____

Name _____   Date _____

11. $\dfrac{130}{-13}$ 11. _____

12. $\dfrac{-96}{-6}$ 12. _____

Evaluate these algebraic expressions.

13. Evaluate $11(x)$ for $x = -7$.   13. _____

14. Evaluate $x \div -8$ for $x = 48$.   14. _____

15. Evaluate $(-x)^3$ for $x = 6$.   15. _____

16. Evaluate $(-x)^2 \div y$ for $x = 4$ and $y = -2$.   16. _____

Practice Set 2.5
The Order of Operations and Applications Involving Integers

Simplify.

1. $-2 + 5 \cdot 6$

    1. _____

2. $-2 - 5(5 - 8)$

    2. _____

3. $2(-5)(7 - 5) - 7$

    3. _____

4. $90 \div (-6) - 19$

    4. _____

5. $3^3 - 8(-1)$

    5. _____

6. $16 - 3(10 - 4^2) + 5$

    6. _____

7. $7 - 2(11 - 3^2) + 4$

    7. _____

8. $(3 - 7)^2 \div (8 - 6)^2$

    8. _____

9. $\dfrac{\left[3^2 + 9(-4)\right]}{\left[6 + (-15)\right]}$

    9. _____

10. $\dfrac{16(-1) - (-4)(-5)}{2\left[-12 \div (-3 - 3)\right]}$

    10. _____

Evaluate these algebraic expressions.

11. Evaluate $4 - x(-3)$ for $x = -6$.   11. _____

12. Evaluate $x^2 + 7(-y) - 3^3$ for $x = -4$ and $y = 10$.   12. _____

13. Evaluate $(x + 2)^2 - (-y - 3)^3$ for $x = -5$ and $y = -2$.   13. _____

Write an expression and evaluate.

14. When Isaac decorated his dining room, he bought a table for   14. _____
     $570, 2 armchairs for $125 each, and 4 chairs for $75 each.
     Write an expression that represents this situation and evaluate it   _____
     to find out how much Isaac spent.

15. Sophia wants to carpet three rooms in her house. Her living room   15. _____
     is 16 feet long and 14 feet wide, her guest room is 13 feet long
     and 12 feet wide, and her bedroom is 11 feet in both directions.   _____
     Write an expression that represents the total area of these three
     rooms in square feet, and then evaluate it.

16. Daisy is an eccentric millionaire who keeps money, stock   16. _____
     certificates, and a record of her debts under her mattress.
     Currently, her mattress contains $10,000 in cash, 100 shares of   _____
     stock valued at $59 each, 300 shares of stock valued at $25 each,
     and an IOU promising to pay her lawyer $5500. Write an
     expression to represent the total value under Daisy's mattress
     and then calculate this amount.

Name _____   Date _____

Practice Set 2.6
Simplifying and Evaluating Algebraic Expressions

Simplify by combining like terms.

1. $-10a - 6a + 3a$ 1. _____

2. $2x + 4x + 6xy$ 2. _____

3. $11 + 3t + 9$ 3. _____

4. $2h + 7k - 12h - 4k$ 4. _____

5. $100 - 3a - 7ab + 5b + 9 + ab$ 5. _____

Evaluate.

6. $3x + 2x$ for $x = -6$ 6. _____

7. $(3x)^2 - 7x$ for $x = -2$ 7. _____

8. $m \cdot n - 4$ for $m = -2$ and $n = 6$ 8. _____

9. $12p + 3p^9 - q^2$ for $p = -1$ and $q = 9$ 9. _____

10. $\dfrac{(a + b^2)}{-2}$ for $a = 15$ and $b = -3$ 10. _____

Simplify by using the distributive property and combining like terms.

11. $3(t + 5)$ 11. _____

12. $-3(2x - 3)$ 12. _____

13. $-7(a + 5) + 7$ 13. _____

14. $-(8i - 3j + 7k) + k$ 14. _____

15. $-3(-4x) - 3 + 8(2x + 9)$ 15. _____

16. $-6(-2a) + 6(a - 3) + 2(-3a) + 19 + a$ 16. _____

Practice Set 3.1
Solving Equations of the Form $x + a = c$ and $x - a = c$

Fill in the blank with the number that gives the desired result.

1. $x + 5 + \underline{\hspace{1cm}} = x + 0 = x$ 1. _____

2. $y - 11 + \underline{\hspace{1cm}} = y + 0 = y$ 2. _____

3. $n + 124 + \underline{\hspace{1cm}} = n + 0 = n$ 3. _____

Solve and check your solution.

4. $x + 7 = 14$ 4. _____

5. $p - 5 = 8$ 5. _____

6. $x + 10 = 25$ 6. _____

7. $m - 13 = 15$ 7. _____

8. $x - 37 = -21$ 8. _____

9. $6 = b - 9$ 9. _____

10. $5x - 4x - 6 = 11$ 10. _____

11. $8n - 32 - 7n = 6$ 11. _____

12. $5 - 11 = a + 2 - 13$ 12. _____

13. $-2 + 4 + x = -3 + 17$ 13. _____

Find the measure of the unknown angle for each pair of supplementary angles ($\angle a + \angle b = 180°$).

14. $\angle a = ?, \angle b = 103°$ 14. _____

15. $\angle a = 82°, \angle b = ?$ 15. _____

Practice Set 3.2
Solving Equations of the Form $ax = c$

Translate each statement into an equation.

1.   Tyrrell has twice as many action figures as Jason.   1. _____

2.   Miriam is seven times older than her daughter Edie.   2. _____

3.   A battleship displaces 10,000 times more water than a yacht.   3. _____

Solve and check your solution.

4. $5m = 15$ 4. _____

5. $2y = 36$ 5. _____

6. $3x = 42$ 6. _____

7. $2x = -26$ 7. _____

8. $-5y = -40$ 8. _____

9. $6y = 2$ 9. _____

10. $4(5x) = 60$ 10. _____

11. $3(2x) = -54$ 11. _____

12. $-14 - 28 = 10x - 3x$ 12. _____

Solve the following problems.

13.   Matthew moved to an apartment building that's 11 times taller   13. _____
      than the house he used to live in. If the apartment building is 253
      feet tall, what's the height of the house?

14.   On Saturday, Gordon catered a birthday party and a wedding.   14. _____
      The wedding had a total of 153 people, which was exactly nine
      times as many people as the birthday party. How many people
      attended the birthday party?

15.   Jenna works at home making flower arrangements. On Monday,   15. _____
      she worked for 420 minutes and on Tuesday she worked for 360
      minutes. All together, she made 39 flower arrangements. If each
      arrangement took her the same amount of time, how many
      minutes did each of them take?

16.   Alicia works as a receptionist for a local appliance store. She   16. _____
      worked four hours on Monday, three hours on Tuesday, and
      seven hours on Wednesday. If her total earnings for that week
      were $116.90, how much does Alicia make per hour?

Name _____   Date _____

Practice Set 3.3
Equations and Geometric Formulas

Solve the following geometry problems.

1. Find the perimeter and area of a rectangle with length of 5 feet   1. _____
    and width of 3 feet.

2. What are the perimeter and area of a square with sides of length   2. _____
    15 yards?

3. Find the area of a parallelogram with a height of 15 inches and   3. _____
    base of 19 inches.

4. What is the height of a parallelogram with a base of 18 feet and   4. _____
    an area of 144 ft^2?

5. The perimeter of a square is 72 meters. What is the measure of   5. _____
    each side of the square?

6. The length of a rectangle is twice its width. If the perimeter of   6. _____
    the rectangle is 42 feet, find the length and width.

7. Find the volume of a rectangular solid with $L = 8$ feet,   7. _____
    $W = 5$ feet, and $H = 2$ feet.

8. A rectangular solid has a volume of 126 feet. If the length of this   8. _____
    solid is 3 feet and the width is 6 feet, what is its height?

35

Solve the following practical problems in geometry.

9. A room measures 18 feet in length and 13 feet in width. How
    many square tiles measuring one foot on a side would you need
    to tile the floor?

9. _____

10. A large field measures 300 yards in one direction and 625 yards
    in the other direction. If you want to walk the entire distance
    around the field, how far do you have to travel?

10. _____

11. Glenn wants to carpet his living room and dining room with
    carpeting that costs $29 per square yard. The living room is
    rectangular and measures four yards by six yards. The dining
    room is a square measuring four yards in one direction. How
    much does Glenn need to spend?

11. _____

12. Tami wants to paint the four walls of a rectangular room with a
    length of 24 feet and a width of 21 feet. If the height of the room
    is 10 feet, how many square feet of wall space does she have to
    cover (ignoring doors and windows)?

12. _____

13. A plastic feed bin is 4 feet long, 2 feet wide, and 2 feet high.
    How many cubic feet of animal feed can be placed in this bin?

13. _____

14. An aquarium has a length of 50 inches, a width of 21 inches, and
    a height of 32 inches. What is the volume of the aquarium?

14. _____

15. Annalisa covered her kitchen floor with 84 linoleum tiles that are
    each one square foot in area. What is the width of her kitchen if
    the length is 12 feet?

15. _____

16. A room holds a volume of 72 cubic meters of air. If the room's
    length is 6 meters and its height is 3 meters, what is the perimeter
    of the floor?

16. _____

Name _____   Date _____

Performing Operations with Exponents

Multiply and write the product in exponent form.

1. $4^3 \cdot 4^2$ 1. _____

2. $2^2 \cdot 2^4 \cdot 2^{11}$ 2. _____

3. $10^3 \cdot 10^5 \cdot 10^9 \cdot 10^0 \cdot 10^1$ 3. _____

Simplify.

4. $x^4 \cdot x^2$ 4. _____

5. $y^3 \cdot y^6 \cdot y^4 \cdot y^7$ 5. _____

6. $(-2x^3)(4x^3)$ 6. _____

7. $(3a)(-2a^4)(-4a^2)$ 7. _____

8. $(5x^2)(3y^2)(-4x^3)$ 8. _____

37

9. $3x(x^2 + 4)$ 9. _____

10. $6y(y^2 - 9x + 2)$ 10. _____

11. $(3a^2 - 7a)(-6a^4)$ 11. _____

12. $2x^5(2x^4 - 5x^3) - 3x^8$ 12. _____

13. $-4x^3(2x + y - 5) + 3x^3y + 4x^4 + y$ 13. _____

Solve these geometry problems.

14. What is the area of a rectangle with a length of $4x$ and a width   14. _____
     of $2x + 7$?

15. What is the area of a parallelogram with a height of $9y - 10$ and   15. _____
     a base of $22y$?

16. What is the volume of a box with a length of 20, a width of $3n$,   16. _____
     and a height of $n - 1$?

Practice Set 4.1
Factoring Whole Numbers

1. List the first 10 prime numbers.   1. _____

Determine if the number is divisible by 2, 3, and/or 5.

2. 20   2. _____

3. 45   3. _____

4. 162   4. _____

5. 1361   5. _____

6. 111,111   6. _____

State whether each number is prime, composite, or neither.

7. 1   7. _____

8. 35   8. _____

9.   43   9. _____

10. 91   10. _____

Express each number as a product of prime factors. Write your answer as powers of prime factors.

11. 44   11. _____

12. 48   12. _____

13. 75   13. _____

14. 98   14. _____

15. 300   15. _____

16. 7200   16. _____

Practice Set 4.2
Understanding Fractions

Write a fraction that represents the situation described.

1. A pizza is divided into eight slices and you eat three of them.   1. _____
    Write a fraction to describe how much of the pizza you ate.

2. Gail belongs to a club with 29 members. She ran for president   2. _____
    and received 15 votes. Write a fraction that describes the
    proportion of votes she received.

3. There are 13 girls and 12 boys at the meeting. Write a fraction   3. _____
    that describes what part of the meeting are girls.

4. Last semester, Mr. Walker's students received 7 A's, 8 B's, 5   4. _____
    C's, 1 D, and 2 F's. Write a fraction to represent the proportion
    of students who received A's.

Change each of the following improper fractions to a mixed number.

5. $\dfrac{8}{3}$ 5. _____

6. $\dfrac{29}{2}$ 6. _____

7. $\dfrac{83}{8}$ 7. _____

8. $\dfrac{44}{11}$ 8. _____

9. $\dfrac{72}{11}$ 9. _____

10. $\dfrac{97}{9}$ 10. _____

Change each of the following mixed numbers to an improper fraction.

11. $3\dfrac{4}{5}$ 11. _____

12. $9\dfrac{7}{9}$ 12. _____

13. $8\dfrac{2}{7}$ 13. _____

14. $10\dfrac{1}{12}$ 14. _____

15. $23\dfrac{4}{5}$ 15. _____

16. $7\dfrac{11}{20}$ 16. _____

Practice Set 4.3
Simplifying Fractional Expressions

Find an equivalent fraction with the given denominator.

1. $\dfrac{1}{2} = \dfrac{?}{12}$

   1. _____

2. $\dfrac{3}{7} = \dfrac{?}{35}$

   2. _____

3. $\dfrac{7}{12} = \dfrac{?}{60}$

   3. _____

4. $\dfrac{7}{9} = \dfrac{?}{36}$

   4. _____

5. $\dfrac{2}{15} = \dfrac{?}{45x}$

   5. _____

6. $\dfrac{5}{8} = \dfrac{?}{64a}$

   6. _____

Simplify.

7. $\dfrac{4}{20}$

   7. _____

8. $\dfrac{22}{33}$

   8. _____

9. $\dfrac{-27}{45}$ 9. _____

10. $\dfrac{-24}{52}$ 10. _____

11. $\dfrac{5x}{10x}$ 11. _____

12. $\dfrac{-45xy}{-72y}$ 12. _____

Solve. Write each answer in simplest form.

13. Lane has 24 plants in his house, 18 of which are in his living
    room. Find a fraction to represent the fractional part of Lane's   13. _____
    plants that are in the living room.

14. On a recent test, Janey answered 45 out of 50 questions
    correctly. What fractional part of the test did she get right?   14. _____

15. A bookshelf contains 120 books, 63 of which are mystery novels.
    What part of these books are not mystery novels?   15. _____

16. Marta is saving up $1250 to buy a racing bicycle. She already
    has $450. Write a fraction to represent the portion that Marta still 16. _____
    needs to save.

Name _____   Date _____

Practice Set 4.4
Simplifying Fractional Expressions with Exponents

Simplify. Assume that all variables in any denominator are nonzero. Leave your answers in exponent form.

1. $\dfrac{3^8}{3^2}$

1. _____

2. $\dfrac{a^9}{a^5}$

2. _____

3. $\dfrac{x^5}{x^{12}}$

3. _____

4. $\dfrac{x^7 y^5}{x^2 y^{11}}$

4. _____

5. $\dfrac{10 a^3 b^7 c^0}{15 a^5 b^5}$

5. _____

6. $\dfrac{7^3 m n^8}{7^5 m^5 n^5}$

6. _____

7. $\dfrac{45 x^0 y^{17} z^4}{63 y^9 z^8}$

7. _____

Write as a product and then simplify. Leave your answer in exponent form.

8. $\left(x^4\right)^3$ 8. _____

9. $\left(b^4\right)^5$ 9. _____

10. $\left(3a^2\right)^3$ 10. _____

11. $\left(2x^3 y^2 z\right)^3$ 11. _____

12. $\left(\dfrac{2}{x}\right)^3$ 12. _____

13. $\left(\dfrac{2}{3}\right)^4$ 13. _____

14. $\left(\dfrac{2a^2 b^3}{c^4}\right)^2$ 14. _____

15. $\left(3y^3 z^2\right)^2 \cdot \left(2y^2\right)^3$ 15. _____

Practice Set 4.5
Ratios and Rates

Write each ratio as a fraction in simplest form.

1. 27:9   1. _____

2. The ratio of 30 to 65   2. _____

3. 55:10   3. _____

4. The ratio of 84 feet to 49 feet   4. _____

5. 60:90   5. _____

6. The ratio of $141 to $69   6. _____

Solve each of the following.

7. Cecelia has five dolls and seven stuffed animals on her bed. 7. _____
    Write a fraction that describes the ratio of dolls to stuffed
    animals.

8. A nursery school has 20 children and four adult supervisors. 8. _____
    Write a fraction that represents the ratio of adults to children.

9. A company has a total staff of 55 workers, which includes the 9. _____
    employees and five managers. How many employees are there
    for each manager?

10. If a car goes 300 miles on a 12-gallon tank of gas, how many miles per gallon does the car get?

10. _____

11. A sign at Ray's Used Uniform Store advertises any six shirts for $51.00. What is the unit price per shirt?

11. _____

12. Nadia ran 16 days last month and covered a total of 88 miles. How many miles did she average each day?

12. _____

13. Cathleen worked six hours and earned $54. Darla worked seven hours and earned $56. Which young woman earns more money per hour?

13. _____

14. If five pounds of chicken costs $10.75 and three pounds of pork costs $8.10, which type of meat costs more per pound?

14. _____

15. A four-ounce can of tuna sells for $1.08. A six-ounce can sells for $1.41, and an eight-ounce can sells for $1.89. Which size can has the least expensive unit price?

15. _____

16. Jason and Kyle compared their travel journals from a recent trip. Jason traveled 2,880 miles and purchased 180 gallons of gasoline. Kyle traveled 3,150 miles and purchased 210 gallons of gasoline. Who owns the vehicle with the better miles per gallon?

16. _____

Practice Set 4.6
Proportions and Applications

Use the equality test for fractions to determine if the fractions are equal.

1. $\dfrac{2}{5} \overset{?}{=} \dfrac{6}{15}$ 1. _____

2. $\dfrac{7}{10} \overset{?}{=} \dfrac{8}{11}$ 2. _____

3. $\dfrac{18}{8} \overset{?}{=} \dfrac{45}{20}$ 3. _____

4. $\dfrac{4}{36} \overset{?}{=} \dfrac{14}{117}$ 4. _____

Find the value of x in each proportion. Check your answer.

5. $\dfrac{3}{4} = \dfrac{x}{12}$ 5. _____

6. $\dfrac{2}{10} = \dfrac{x}{15}$ 6. _____

7. $\dfrac{x}{32} = \dfrac{5}{16}$ 7. _____

8. $\dfrac{x}{42} = \dfrac{1}{14}$ 8. _____

9. $\dfrac{4}{13} = \dfrac{52}{x}$

9. _____

10. $\dfrac{2}{19} = \dfrac{10}{x}$

10. _____

Solve each of the following.

11. A package of three T-shirts costs $11. How much will 12 T-shirts cost?

11. _____

12. Candace reads three books every four weeks. At that rate, how many books will she read in 20 weeks?

12. _____

13. If Tamara takes 15 minutes to stock three shelves of merchandise, how long will it take her to stock 14 shelves?

13. _____

14. On a map, 5 miles is represented by 1 inch. If you need to travel $3\dfrac{1}{2}$ inches on the map, how far are you really traveling?

14. _____

15. During a long camping trip, Noreen covered 117 kilometers in 13 days. At that rate, how far can she expect to travel in half a day?

15. _____

Practice Set 5.1
Multiplying and Dividing Fractional Expressions

Multiply. Be sure your answer is simplified.

1. $\dfrac{2}{5} \cdot \dfrac{1}{9}$ 1. _____

2. $\dfrac{5}{8} \cdot \dfrac{3}{7}$ 2. _____

3. $\dfrac{3}{5} \cdot \dfrac{11}{20}$ 3. _____

4. $\dfrac{-1}{6} \cdot \dfrac{18}{19}$ 4. _____

5. $\dfrac{11}{14} \cdot \dfrac{26}{22}$ 5. _____

6. $27 \cdot \dfrac{2}{9}$ 6. _____

7. $\dfrac{7x}{3} \cdot \dfrac{3x}{2}$ 7. _____

8. $\left(\dfrac{-3x}{4}\right) \cdot \left(\dfrac{6}{7x}\right) \cdot \left(\dfrac{2}{-5x}\right)$ 8. _____

Name _____   Date _____

Divide. Be sure your answer is simplified.

9. $\dfrac{3}{5} \div \dfrac{5}{6}$

                                                                             9. _____

10. $\dfrac{5}{9} \div \dfrac{7}{3}$

                                                                             10. _____

11. $\dfrac{2}{3} \div \left(\dfrac{-1}{3}\right)$

                                                                             11. _____

12. $\left(\dfrac{-5}{12}\right) \div \dfrac{25}{36}$

                                                                             12. _____

13. $\left(\dfrac{5x^2}{11}\right) \div \left(\dfrac{30}{77x}\right)$

                                                                             13. _____

14. $15 \div \dfrac{3}{7}$

                                                                             14. _____

15. $27x^4 \div \dfrac{9}{4x^3}$

                                                                             15. _____

Practice Set 5.2
Multiples and Least Common Multiples of Algebraic Expressions

Find the LCM of each group of expressions.

1. 7 and 14

1. _____

2. 13 and 26

2. _____

3. 12 and 18

3. _____

4. 15 and 20

4. _____

5. 20 and 45

5. _____

6. 6, 12, and 24

6. _____

7. 3, 6, and 21

7. _____

8. $27x$ and $36x$

8. _____

9. $16x$ and $32x^2$

9. _____

10. $27x^3$ and $15x$ 10. _____

11. $25x^3$ and $30x^4$ 11. _____

12. $16x$, $70x^2$, and $7x^3$ 12. _____

13.   The Admissions Office of the college conducts campus tours.   13. _____
      Amy, a campus tour guide, gives a 30-minute tour of the campus
      while Angela's tour runs only 20 minutes. They each take a 5-
      minute break between each tour. If they begin tours at 9 A.M.,
      what is the next time that both tours will start at the same time?

14.   Francesco and Marion are walking laps around a track.   14. _____
      Francesco walks 1 lap every 6 minutes while Marion walks 1 lap
      every 8 minutes. If Francesco and Marion begin their walk at the
      same time and location on the track, in how many minutes will
      they meet to begin their next lap together?

15.   Two security guards patrol a museum each night. Both security   15. _____
      guards begin their patrol from the same point. The security guard
      patrolling the north wing takes 21 minutes to complete his
      rounds, while the security guard patrolling the south wing takes
      18 minutes to complete his rounds. If both security guards leave
      the same point at 10:00 P.M., at what time will they cross paths?

Name _____   Date _____

Practice Set 5.3
Adding and Subtracting Fractional Expressions

Perform the operation indicated. Be sure to simplify your answer.

1. $\dfrac{2}{11} + \dfrac{5}{11}$ 1. _____

2. $\dfrac{4}{15} + \left(\dfrac{-2}{15}\right)$ 2. _____

3. $\dfrac{-3}{16} + \left(\dfrac{-5}{16}\right)$ 3. _____

4. $\dfrac{5}{7} + \dfrac{2}{7}$ 4. _____

5. $\dfrac{5}{7} + \dfrac{23}{7}$ 5. _____

6. $\dfrac{22}{y} + \dfrac{5}{y}$ 6. _____

7. $\dfrac{1}{2} + \dfrac{1}{5}$ 7. _____

8. $\dfrac{3}{10} + \dfrac{4}{25}$ 8. _____

9. $\dfrac{-5}{8}+\dfrac{1}{6}$

9. _____

10. $\dfrac{2}{7}+\dfrac{5}{9}$

10. _____

11. $\dfrac{1}{4x}+\dfrac{7}{30x}$

11. _____

12. $\dfrac{-3x}{14}-\left(\dfrac{-8x}{21}\right)$

12. _____

13. $\dfrac{5}{a}+\dfrac{2}{b}$

13. _____

14. Planning a cookout, Dreana purchased $6\dfrac{1}{2}$ pounds of hamburger, 3 pounds of hot dogs, and $8\dfrac{1}{3}$ pounds of chicken. How much meat did Dreana purchase?

14. _____

15. Tran and Alex were collecting canned food as part of their class food drive. Tran collected $\dfrac{3}{8}$ of the class total and Alex collected $\dfrac{1}{5}$ of the class total. What part of the total number of cans did Tran and Alex collect together?

15. _____

Practice Set 5.4
Operations with Mixed Numbers

Add or subtract. Simplify all answers. Express as a mixed number.

1. $13\dfrac{1}{3} + 2\dfrac{1}{3}$

   1. _____

2. $8\dfrac{1}{7} + 5\dfrac{4}{7}$

   2. _____

3. $10\dfrac{6}{7} - 6\dfrac{5}{7}$

   3. _____

4. $12\dfrac{5}{9} - 4\dfrac{1}{6}$

   4. _____

5. $9\dfrac{3}{5} + 7\dfrac{4}{5}$

   5. _____

6. $5\dfrac{6}{7} + 8\dfrac{11}{14}$

   6. _____

7. $25\dfrac{7}{11} - 13\dfrac{10}{11}$

   7. _____

8. $14\dfrac{3}{8} - 5\dfrac{9}{16}$

   8. _____

Multiply or divide and simplify your answer.

9. $2\frac{1}{4} \cdot 3\frac{1}{3}$

9.   _____

10. $\left(\frac{-2}{7}\right) \cdot 3\frac{5}{8}$

10. _____

11. $2\frac{3}{5} \cdot 4\frac{5}{8}$

11. _____

12. $6\frac{1}{4} \div 2\frac{1}{2}$

12. _____

13. $-2\frac{3}{7} \div \frac{3}{14}$

13. _____

14. $7\frac{1}{2} \div (-3)$

14. _____

15. $(-5) \div \left(-3\frac{2}{7}\right)$

15. _____

Name _____   Date _____

Practice Set 5.5
Order of Operations and Complex Fractions

Simplify.

1. $4 \cdot 5^2 - \dfrac{47}{3}$

    1. _____

2. $\left(\dfrac{3}{5}\right)^2 \cdot \dfrac{1}{3}$

    2. _____

3. $\left(\dfrac{3}{5}\right)^3 \cdot \left(\dfrac{1}{3}\right)^3$

    3. _____

4. $\dfrac{3}{2} + \left(\dfrac{4}{3}\right)^2 - \dfrac{5}{9}$

    4. _____

5. $\left(\dfrac{4}{5} + \dfrac{4}{15}\right) \cdot \dfrac{5}{6}$

    5. _____

6. $\dfrac{3}{8} \cdot \left(\dfrac{1}{4} + \dfrac{1}{2}\right) \cdot \dfrac{32}{3}$

    6. _____

7. $\left(\dfrac{2}{5} - \dfrac{3}{10}\right)\left(\dfrac{2}{5} + \dfrac{3}{10}\right)$

    7. _____

8. $\left(\dfrac{3}{8}\right)^2 \left(\dfrac{3}{8} + \dfrac{5}{8}\right)$

    8. _____

9. $\dfrac{7+(-3)^2}{\dfrac{8}{10}}$ 9. _____

10. $\dfrac{\dfrac{3}{5}}{4^2-10}$ 10. _____

11. $\dfrac{\dfrac{5}{7}}{\dfrac{8}{14}}$ 11. _____

12. $\dfrac{\dfrac{x^2}{3}}{\dfrac{x}{18}}$ 12. _____

13. $\dfrac{\dfrac{3}{8}+\dfrac{1}{4}}{\dfrac{3}{5}+\dfrac{1}{10}}$ 13. _____

14. $\dfrac{\dfrac{7}{8}-\dfrac{1}{4}}{\dfrac{5}{6}-\dfrac{1}{3}}$ 14. _____

15. $\dfrac{\dfrac{16}{30}-\dfrac{2}{15}}{\dfrac{14}{40}+\dfrac{9}{20}}$ 15. _____

Practice Set 5.6
Solving Applied Problems Involving Fractions

When solving the following application problems, you may want to use a Mathematics Blueprint for Problem Solving to help organize your work.

1. Chris and Alicia traveled 160 miles in $3\frac{1}{5}$ hours. On the average, how many miles per hour did they drive?

   1. _____

2. A recipe called for $2\frac{1}{3}$ cups of milk, $\frac{1}{4}$ cup of flour, and 4 tablespoons of butter. This recipe will make a sauce for 8 servings. How much of each ingredient will you need to make 6 servings?

   2. _____

3. A piece of rope $27\frac{1}{2}$ yards long is used to tie 5 boxes. How long is each piece of rope for each box if all the rope is used?

   3. _____

4. Amy and Alex have 55 feet of ribbon. How much ribbon will be left if they use $\frac{1}{5}$ of the ribbon to decorate their raffle table and they use $\frac{5}{7}$ of the ribbon to hang nametags?

   4. _____

5. Allison and Tyler are planning a cookout. They invited 27 people (including themselves) to the cookout. If they estimate that each person can eat $\frac{1}{3}$ pound of meat, $\frac{3}{5}$ pound of potato salad, and $\frac{1}{2}$ pound of fruit, how much meat, salad, and fruit must they order?

   5. _____

6. Colin needs to drill a hole through 5 pieces of wood, each of which is $4\frac{3}{4}$ inches wide, and then insert a metal rod through the hole. What is the minimum length of the metal rod that must be inserted through the hole?

   6. _____

7. A machine that cuts metal is calibrated to cut strips $1\frac{3}{8}$ inches   7. _____
    long from a piece of metal that is 24 inches long. How many
    strips will the machine cut from this one piece?

8. Peter had $18\frac{3}{4}$ gallons of gasoline in his car before he began his   8. _____
    vacation. When he arrived at his destination, he only had $7\frac{1}{3}$
    gallons left. How much gasoline did he use on his trip?

9. Plastic tubing is sold in 40-foot bundles. To complete an order,   9. _____
    Kieran needs to cut 48 sections of plastic tubing that are each
    $3\frac{5}{8}$ feet long. How many bundles of plastic tubing will Kieran
    need to purchase to complete the order?

10. Paula wins $2500 in a local lottery. She spends $\frac{1}{8}$ of her money   10. _____
    on a new computer monitor and $\frac{3}{5}$ of her money to pay off some
    bills. The remainder is deposited into her checking account. How
    much money from her lottery winnings will she put into her
    checking account?

11. The instructions for making a concrete driveway require $13\frac{1}{3}$ 11. _____
    parts cement, 40 parts sand, and 90 parts aggregate. How many
    parts cement are required to make 8 concrete driveways?

12. A border of bricks $1\frac{1}{4}$ feet wide surrounds a patio. The   12. _____
    combined patio and border measure 25 feet by 30 feet. What are
    the dimensions of the patio?

Practice Set 5.7

Solving Equations of the Form $\dfrac{x}{a} = c$

Solve and check your solutions.

1. $\dfrac{x}{3} = 12$ 1. _____

2. $\dfrac{x}{5} = 15$ 2. _____

3. $\dfrac{x}{12} = 7$ 3. _____

4. $\dfrac{x}{8} = 22$ 4. _____

5. $-3 = \dfrac{x}{-32}$ 5. _____

6. $-14 = \dfrac{x}{4}$ 6. _____

7. $\dfrac{x}{5^2} = 11$ 7. _____

8. $\dfrac{x}{-9} = -27 + 12$ 8. _____

9. $\dfrac{x}{5^3} = -2$ 9. _____

10. $\dfrac{x}{-6} = 14 + (-16)$ 10. _____

11. $\dfrac{1}{3}x = 11$ 11. _____

12. $-\dfrac{1}{7}x = 5$ 12. _____

13. $\dfrac{1}{8}x = -14$ 13. _____

14. $\dfrac{2}{3}x = 6$ 14. _____

15. $\dfrac{5}{7}x = 25$ 15. _____

Practice Set 6.1
Adding and Subtracting Polynomials

Identify the terms of each polynomial.

1. $3x^2 + 4xy + 5y^2$

    1. _____

2. $5a^4 - 3a^3 + 2a^2 - 7a + 1$

    2. _____

Perform the operations indicated.

3. $(6x - 2) + (-3x - 4)$

    3. _____

4. $(4x + 1) + (-x - 9)$

    4. _____

5. $(3y^2 - 2y - 5) + (4y^2 - 6y + 4)$

    5. _____

6. $(-3c^2 + 2c + 5) + (-5c^2 - 8)$

    6. _____

Simplify.

7. $-(6x + 10)$

    7. _____

8. $-(-4a^4 - 6a^2 + 9)$

    8. _____

Perform the operations indicated.

9. $(10y + 3) - (-2y + 4)$ 9. _____

10. $(-7y - 3) - (-6y - 4)$ 10. _____

11. $(2a + 12) - (5a - 16)$ 11. _____

12. $(3y^3 - 2y^2 + 5) - (-2y^3 - 5y^2 - y)$ 12. _____

13. $2y - (-2y^3 - 4y - 3) + (-y^2 - 6y - 9)$ 13. _____

14. $7x - 3(x + 4) - (-5x - 9) - (4x + 3)$ 14. _____

15. $-(-5x^5 - 3x^3 + x + 4) - (3x^4 + 2x^2 - 11)$ 15. _____

Name _____   Date _____

Practice Set 6.2
Multiplying Polynomials

Use the distributive property to multiply.

1. $5(3x^2 + 2x - 4)$ 1. _____

2. $-2x(2x^2 - 9x - 5)$ 2. _____

3. $(-4x^2)(5x^3 + 2x^2 - 5)$ 3. _____

4. $(-5x^6)(3x^2 - 7x - 10)$ 4. _____

5. $(2x - 1)(x^2 + 3x + 1)$ 5. _____

6. $(2x - 1)(-3x^2 - 7x + 1)$ 6. _____

Use FOIL to multiply.

7. $(a + 3)(a + 2)$ 7. _____

8. $(x + 7)(x - 5)$ 8. _____

67

9. $(y - 5)(y - 12)$ 9. _____

10. $(5y - 3)(2y - 7)$ 10. _____

11. $(3x + 2)(2x - 4)$ 11. _____

12. $(-x - 9)(-2x + 9)$ 12. _____

Simplify.

13. $-3x(2x^2 - 2x + 1) + (x + 3)(x - 5)$ 13. _____

14. $(x + 4)(x - 2) + 5(3x - 2)$ 14. _____

15. $-2x(x^2 + 3x + 1) + (x - 3)(x + 4)$ 15. _____

Name _____   Date _____

Practice Set 6.3
Translating from English to Algebra

1. The length of a swimming pool is three times the   1. _____
    width. Define the variable expressions for the
    swimming pool's length and width.   _____

2. The price of a new vehicle is $10,100 more than the   2. _____
    price of a used vehicle of the same model. Define the
    variable expressions for the cost of each vehicle.   _____

3. Raymond's monthly salary is $125 more than Anne   3. _____
    Marie's monthly salary. Define the variable
    expressions for the monthly salaries of Raymond and   _____
    Anne Marie.

4. The width of a rectangle is 15 inches shorter than   4. _____
    twice the length. Define the variable expressions for
    the length and width of the rectangle.   _____

5. Sharon is 3 inches taller than Andy. Kari is 5 inches 5. _____
    taller than Andy. Define the variable expressions for
    the heights of Sharon, Kari, and Andy.   _____

6. The height of a flagpole is $\frac{1}{3}$ the height of a building.   6. _____
    Define the variable expressions for the heights of the
    flagpole and building.   _____

7. Chris won twice as much playing poker as Ray. Jay   7. _____
    won $35 less than Ray. Define the variable
    expressions for the winnings of Chris, Ray, and Jay.   _____

8. The second side of a triangle is 4 inches longer than the first. The third side is 6 inches shorter than three times the first. Define the variable expressions for the length of each side of the triangle.

8. _____

   _____

9. Pond B contains 40 fewer fish than Pond A. Pond C contains 15 more fish than two times the number of fish in Pond A. Define the variable expressions for the number of fish in each pond.

9. _____

   _____

10. Ally has 5 more coins in her pocket than Ayla. Jess has 8 fewer coins in her pocket than Ayla.
    a. Define the variable expressions for the numbers of coins in Ally's, Ayla's, and Jess's pockets.
    b. Write the following phrase using math symbols: The number of coins in Ally's pocket plus the number of coins in Jess's pocket minus the number of coins in Ayla's pocket.
    c. Simplify the expression from part b.

10. a._____

    _____

    b. _____

    c. _____

11. Brandon has 350 more hockey cards in his collection than Sheldon. Trevor has 180 fewer hockey cards in his collection than Sheldon.
    a. Define the variable expressions for the numbers of hockey cards in Brandon's, Sheldon's, and Trevor's collections.
    b. Write the following phrase using math symbols: The number of hockey cards in Brandon's collection plus the number of hockey cards in Trevor's collection minus the number of hockey cards in Sheldon's collection.
    c. Simplify the expression from part b.

11. a.._____

    _____

    b. _____

    c. _____

Practice Set 6.4
Factoring Using the Greatest Common Factor

Find the GCF for each expression.

1. $ab^3 + a^2b^2$ 1. _____

2. $x^4y^5 - x^3y^6$ 2. _____

3. $m^2n^3 + a^2b^2$ 3. _____

Factor. Check by multiplying.

4. $4a + 8$ 4. _____

5. $7x + 7$ 5. _____

6. $3x + 30$ 6. _____

7. $3x - 9$ 7. _____

8. $8m - 24$ 8. _____

9. $5a - 45$

10. $3m - 9n$ 10. _____

11. $6a + 8b + 14$ 11. _____

12. $30x - 12y + 18$ 12. _____

13. $8y^2 - 6y$ 13. _____

14. $12a^2b - 20ab^2$ 14. _____

15. $27x^3y^2 - 34a^3b^3$ 15. _____

Name _____   Date _____

Practice Set 7.1
Solving Equations Using One Principle of Equality

Solve and check your solution.

1. $x - 5 = -12$ 1. _____

2. $x + 11 = 23$ 2. _____

3. $-31 = a + 7$ 3. _____

4. $-12 = b - 6$ 4. _____

5. $x - 7 = 11$ 5. _____

6. $10 - 18 = 8 + x - 3$ 6. _____

7. $-4 - 2 = 6y + 2 - 5y - 3$ 7. _____

8. $-12 - 3 = 6x - 10 - 5x + 3$ 8. _____

9. $-6 + 4 = 10x - 11 - 9x - 2$

9. _____

10. $\dfrac{y}{-6} = 5 + 3^2$

10. _____

11. $\dfrac{a}{-4} = 12 - 4^2$

11. _____

12. $\dfrac{2}{3}x = 2^3 - 6$

12. _____

13. $\dfrac{10}{-2} = 4(-3x) + 6x$

13. _____

14. $-x = 6$

14. _____

15. $-x = 33$

15. _____

Name _____   Date _____

Solve and check your solution.

1. $3x + 23 = 26$ 1. _____

2. $3x + 8 = 38$ 2. _____

3. $20x + 19 = 19$ 3. _____

4. $9x - 1 = 26$ 4. _____

5. $10 - x = 12$ 5. _____

6. $15 = 5y + 20$ 6. _____

7. $11 = 2x - 4$ 7. _____

8. $-24 = 2x - 2$ 8. _____

9. $14x + 24 = 12x + 28$ 9. _____

10. $x - 24 = 64 - 10x$ 10. _____

11. $8x + 1 - 3x - 7 = 7$ 11. _____

12. $-2 + 2x + 4 = 12x - 28$ 12. _____

13. $-7 + 12x - 12 = 7x - 3$ 13. _____

14. $-3x + 5 + 4x - 3 = 15 - 27$ 14. _____

15. $-9x + 5 + 7x = -3x + 10$ 15. _____

Practice Set 7.3
Solving Equations with Parentheses

Solve and check your solution.

1. $5(3b - 1) = 40$ 1. _____

2. $-3(2y - 1) = 45$ 2. _____

3. $-3(a - 1) = 21$ 3. _____

4. $-(3x - 1) + 5x = 3$ 4. _____

5. $-2(3x - 1) + 2x = 6$ 5. _____

6. $5(y - 2) - 6y = 12$ 6. _____

7. $-5(2x + 1) + 4x = 7$ 7. _____

8. $-4(x + 3) + 5(x + 6) = 39$ 8. _____

9. $-5(y + 1) + 4(y - 5) = 11$ 9. _____

10. $3(x + 1) + 4(x + 2) = 32$ 10. _____

11. $-4(y + 6) + 2(y + 2) = 18$ 11. _____

12. $(2y^2 - 4y + 1) - (2y^2 + 4) = 29$ 12. _____

13. $(6y^2 + 4y + 1) - (6y^2 + 5) = 20$ 13. _____

14. $(6x^2 - 3x + 1) - (6x^2 - 6) = 2x + 12$ 14. _____

15. $(y^2 + 3y - 4) - (y^2 + 4y - 6) = 6$ 15. _____

Name _____   Date _____

Solve and check your solution.

1. $\dfrac{x}{2} + \dfrac{x}{3} = 15$ 1. _____

2. $\dfrac{x}{2} + \dfrac{x}{5} = 7$ 2. _____

3. $\dfrac{x}{4} + \dfrac{x}{5} = 3$ 3. _____

4. $\dfrac{x}{3} - \dfrac{1}{3} = -6$ 4. _____

5. $\dfrac{x}{5} - \dfrac{x}{3} = 2$ 5. _____

6. $\dfrac{x}{8} - \dfrac{x}{9} = 7$ 6. _____

7. $3x - \dfrac{1}{2} = \dfrac{1}{4}$ 7. _____

8. $6x + \dfrac{1}{2} = \dfrac{2}{3}$ 8. _____

9. $-2x + \dfrac{3}{4} = \dfrac{1}{6}$

9. _____

10. $-4x - \dfrac{2}{3} = \dfrac{5}{6}$

10. _____

11. $\dfrac{x}{2} + x = 9$

11. _____

12. $\dfrac{x}{10} + x = 7$

12. _____

13. $\dfrac{x}{3} - 2x = 4$

13. _____

14. $\dfrac{x}{4} - 5x = 19$

14. _____

15. $\dfrac{x}{5} - 3x = 4$

15. _____

Name _____   Date _____

Practice Set 7.5
Using Equations to Solve Applied Problems

Use the appropriate formula to solve Exercises 1-4.

1. Find x if the perimeter of the rectangle is 56 meters.   1.

$x + 10$

5 m

2. Find x if the perimeter of the rectangle is 62 yards.   2.

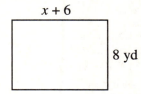

$x + 6$

8 yd

3. Find x if the perimeter of the triangle is 54 feet.   3.

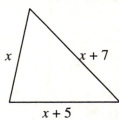

x $x + 7$

$x + 5$

4. Find x if the perimeter of the triangle is 22 inches.   4.

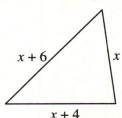

$x + 6$ x

$x + 4$

Name _____   Date _____

Solve and check your solution.

5. June wants to enlarge her rectangular garden area from
    width = 3 feet and length = 4 feet. She plans to increase the
    length by x feet so the new perimeter is 24 feet.
    a. How much should the length be increased?
    b. What will be the length of the enlarged garden?

    5. a._____
       b._____

6. Sally wants to enlarge her rectangular garden area from
    width = 3 feet and length = 4 feet. She plans to increase the width
    by x feet so the new perimeter is 18 feet.
    a. How much should the width be increased?
    b. What will be the dimensions of the enlarged garden?

    6. a._____
       b._____

7. An aerobics room needs to be enlarged. The original room is 56
    feet long and 22 feet wide. The plan calls for the width to be
    increased by x feet, and the new perimeter will be 170 feet. How
    much should the width of the aerobics room be increased?

    7. _____

8. Bob, an experienced assistant store manager, earns $2500 more
    annually than Carl, a new assistant store manager. The sum of
    Bob's annual salary and Carl's annual salary is $41,500. How
    much does each earn annually?

    8. _____

9. John spends a fixed amount on fast food each month. Kevin
    spends $50 more each month on fast food than John does. The
    sum of John's monthly fast food expenditures and Kevin's
    monthly fast food expenditures is $450. How much does each
    spend monthly on fast food?

    9. _____

10. Sue, an executive, earns twice as much as her secretary Sam
    annually. The sum of Sue's annual salary and Sam's annual
    salary is $93,000. How much does each earn annually?

    10. _____

Name _____   Date _____

Practice Set 8.1
Understanding Decimal Fractions

Write a word name for each decimal.

1.   3.48   1. _____

                                                  _____

2.   0.783   2. _____

                                                  _____

Write the word name for the amount on each check.

3.   A check written to John Rodriguez for   3. _____
     $22.43
                                                  _____

4.   A check written to Sarah Clark for   4. _____
     $392.05
                                                  _____

Write in fractional notation. Do not simplify.

5.   0.4   5. _____

6.   2.78   6. _____

7.   5.327   7. _____

8.   121.0025   8. _____

Name _____   Date _____

Write each fraction as a decimal.

9. $\dfrac{3}{10}$ 9. _____

10. $\dfrac{73}{100}$ 10. _____

11. $19\dfrac{243}{1000}$ 11. _____

Replace the ? with < or >.

12.   0.35 ? 0.355   12. _____

13.   0.261 ? 0.26   13. _____

Round to the appropriate decimal place.

14.   Round 3.1415 to the nearest hundredth.   14. _____

15.   Round 3.1415 to the nearest thousandth.   15. _____

Practice Set 8.2
Adding and Subtracting Decimal Expressions

Add.

1.   3.6
        +7.2

1. _____

2.   0. 2 3
      + 2. 6 5

2. _____

3.   0. 4 7
      + 9. 5 7

3. _____

4.   2 9. 3
      + 1 2. 0 4 7

4. _____

5.   7.334 + 21.04

5. _____

Subtract.

6.   1 4.7
      −6.3

6. _____

7.   1 2.2
      −3.6

7. _____

8.   2 0. 0 0
        − 0. 7 3

8. _____

9.   7 9. 4 9 8
     − 6. 2 1 7

9. _____

10.   1 0 0. 0 4 3
     − 9 8. 9 7 5

10. _____

11. $-12.6 - (-6.2)$

11. _____

12. $-30.79 - (-7.91)$

12. _____

Combine like terms.

13. $4.3x + 9.6x - 2.1y$

13. _____

14. $10.02x - 7.39x + 3.4y$

14. _____

Evaluate for the given value.

15. $x - 0.97$ for $x = 13.23$

15. _____

Name _____   Date _____

Practice Set 8.3
Multiplying and Dividing Decimal Expressions

Multiply.

1. 0.02 × 0.09   1. _____

2. 0.06 × 0.8   2. _____

3. 0.04 × 0.03   3. _____

4. 0.076 × 2.3   4. _____

5. (15.23)(–3)   5. _____

6. (–2)(0.0083)   6. _____

7. 2.3681 × 1000   7. _____

8. 798.32 × 10^3 8. _____

Name _____   Date _____

Divide.

9.   −0.21 ÷ 15   9. _____

10.   143.676 ÷ 2.763   10. _____

11.   13.5525 ÷ 4.17   11. _____

12.   0.7 ÷ 1.4   12. _____

Write as a decimal. Round to the nearest hundredth when necessary. If a repeating decimal is obtained, use proper notation such as $0.\overline{3}$.

13. $\dfrac{17}{7}$ 13. _____

14. $\dfrac{7}{11}$ 14. _____

15. $11\dfrac{4}{15}$ 15. _____

Name _____   Date _____

Practice Set 8.4
Solving Equations and Applied Problems Involving Decimals

Solve.

1. $x + 2.4 = 8.5$

                                                     1. _____

2. $x - 3.6 = 7.2$

                                                     2. _____

3. $y - 2 = 7.8$

                                                     3. _____

4. $-0.4x = -16$

                                                     4. _____

5. $-2.2x = -13.2$

                                                     5. _____

6. $-3.45x = 3.864$

                                                     6. _____

7. $5(3x - 0.4) = 4x - 2$

                                                     7. _____

8. $3(x - 1.3) = 9.3$

                                                     8. _____

9. $4x + 6.2 = -9$ 9. _____

10. $0.5x + 0.6 = -0.4$ 10. _____

11. $5x + 11.5 = x - 21$ 11. _____

12. $0.4x - 1.5 = 3.6$ 12. _____

13. $0.07x + 0.27 = 1.46$ 13. _____

Solve the following applied problems.

14. Mike has a collection of coins which consists of dimes and   14. _____
     quarters. The value of these coins is $4.25. If the number of
     quarters in his collection is 3 times the number of dimes, how
     many of each coin does he have?

15. Joseph went shopping at Sami's CD/DVD Store. He purchased 2   15. _____
     CDs at $13.99 each, 1 CD set for $18.99, and 2 DVDs for $19.99
     each (sales tax was included). When reaching the cashier, Joseph
     had only $100 in his wallet. Did he have enough money?

Practice Set 8.5
Estimating with Percents

For the number 56, estimate the following.

1. 10%   1. _____

2. 20%   2. _____

3. 50%   3. _____

For the number 407, estimate the following.

4. 10%   4. _____

5. 30%   5. _____

6. 5%   6. _____

For the number 3006, estimate the following.

7. 1%   7. _____

8. 2%   8. _____

9.   8%   9. _____

For the number 420,070, estimate the following.

10.   15%   10. _____

11.   20%   11. _____

12.   4%   12. _____

Solve the following applications.

13.   Tom would like to leave a 20% tip for a waiter. If the total bill at   13. _____
      the restaurant was $30.61, estimate the tip Tom should leave.

14.   Anastasia must pay a 6% late fee on her electric bill. If her   14. _____
      electric bill is $148.50, estimate the amount of the late fee.

15.   Matt and Andrea paid the realtor 7% commission on the sale   15. _____
      price of their home. If they sold their home for $205,000,
      estimate how much commission was paid to the realtor.

Name _____   Date _____

Practice Set 8.6
Percents

1. 47 out of 100 students in the graduating high school class went to   1. _____
    college. What percentage of the class went to college?

2. 65 out of 100 people attending a conference owned a cell phone.   2. _____
    What percentage of the people owned a cell phone?

3. Last year 100 students attended the annual baseball game. This   3. _____
    year 123 students attended. Write this year's attendance as a
    percent of last year's.

4. 0.6 out of 100 mL of solution is not water. What percentage of   4. _____
    the solution is not water?

Write the following percents as a decimal.

5. 37%   5. _____

6. 97.65%   6. _____

7. 0.017%   7. _____

8. 245%   8. _____

Write the following decimals as a percent.

9.   0.385   9. _____

10. 1.10   10. _____

11. 0.07   11. _____

Write each fraction as a percent. Round to the nearest hundredth of a percent.

12. $\dfrac{1}{5}$ 12. _____

13. $5\dfrac{1}{3}$ 13. _____

Write each percent as a fraction.

14. Write 2.5% as a fraction.   14. _____

15. Write $\dfrac{1}{8}$% as a fraction.   15. _____

Practice Set 8.7
Solving Percent Problems Using Equations

Translate into an equation and solve.

1.   What is 10% of 600?   1. _____

2.   What is 25% of 80?   2. _____

3.   What is 32% of 1210?   3. _____

4.   What is 75% of 120?   4. _____

5.   What is 83% of 155?   5. _____

6.   What is 235% of 120?   6. _____

Translate into an equation and solve.

7.   23 is 1% of what number?   7. _____

8.   240 is 80% of what number?   8. _____

9.   17.2 is 20% of what number?   9. _____

Translate into an equation and solve.

10.   6 is what percent of 25?   10. _____

11.   12 is what percent of 80?   11. _____

12.   280 is what percent of 70?   12. _____

Solve each applied problem.

13.   Last year, Raymond bought a share of stock for $147.50.   13. _____
      Raymond was paid a dividend of $8.85. Determine what percent
      of the stock price is the dividend.

14.   Julie's bill for dinner at the Seafood Hut was $34.75. How much   14. _____
      should she leave for a 15% tip? Round the amount to the nearest
      cent.

15.   Carl received 40 out of 50 points on a quiz. What percent of the   15. _____
      total points did Carl receive on the quiz?

Practice Set 8.8
Solving Percent Problems Using Proportions

Identify the percent *p*, base *b*, and amount *a*. Do not solve for the unknown.

		Percent (*p*)	Base (*b*)	Amount (*a*)
1.	24% of 37 is 9.			
2.	36 is 120% of 30.			
3.	What percent of 350 is 35?			
4.	39.2 is what percent of 56?			
5.	What is 88% of 198?			

In Exercises 6-8, the amount *a* is not known.

6.   Find 35% of 280.   6. _____

7.   210% of 30 is what?   7. _____

8.   Find 0.02% of 950.   8. _____

Name _____

In Exercises 9-11, the base b is not known.

9.   92% of what is 276?   9. _____

10. 40% of what is 38?   10. _____

11. 1800 is 225% of what?   11. _____

In Exercises 12-14, the percent p is not known.

12. 80 is what percent of 400?   12. _____

13. 96 is what percent of 240?   13. _____

14. What percent of 300 is 36?   14. _____

Solve using the percent proportion.

15. Jim puts $165 of his monthly salary of $1650 into a savings plan.   15. _____
     What percent of Jim's salary does he withhold for savings?

Practice Set 8.9
Solving Applied Problems Involving Percents

Solve. If necessary, round your answer to the nearest hundredth.

1. Jane is an art dealer who earns 15% on each work of art she sells.   1. _____
    If she earned $1200 in commissions this month, what were her
    total sales for the month?

2. You must pay annual property tax of 1.25% of the value of your   2. _____
    home. If your home is worth $140,000, how much must you pay
    in taxes each year?

3. Dempsey's Used Car Sales pays its sales personnel an 11%   3. _____
    commission rate based on the price of the car. Last month, Jack
    sold a total of $76,000 worth of cars. What was Jack's
    commission last month?

4. Bill works as a phone solicitor and is paid an 8% commission on   4. _____
    the amount of sales he makes. If Bill earned $280 in
    commissions last week, what were his total sales for the week?

5. A recent sale on pocketbooks advertised 20% off the original   5. _____
    price. Joan and Kathy found a pocketbook with an original price
    of $36.50. With the advertised sale, how much did they pay?

6. Elizabeth's salary last year was $43,500. If she gets a 5.5% pay raise, what is her new salary?

6. _____

7. Dan puts $2400 into a certificate of deposit (CD) account at a simple interest rate of 6% per year.
    a. How much interest will Dan earn in 1 year?
    b. How much money will Dan have in his account at the end of the year?

7. a. _____

   b. _____

8. Neil Paquette sells electronic surveillance equipment. He earns a base salary of $1500 per month plus a commission of 12% of his total sales.
    a. If Neil sold $27,000 in surveillance equipment during the month of February, what was his commission?
    b. What were Neil's total earnings for the month of February?

8. a. _____

   b. _____

9. On the first geometry exam, James earned a score of 75. He studied harder for the second geometry exam and his score on the second exam was 20% percent higher than his score on the first exam. What was his score on the second exam?

9. _____

10. Marta borrows $7500 at a simple annual interest rate of 8%. Six months later, she repays the loan. How much interest does she pay on the loan?

10. _____

Name _____   Date _____

Practice Set 9.1
Interpreting and Constructing Graphs

The double-bar graph below compares the number of computer systems sold by two sales associates over four years. Use this graph to answer the following questions.

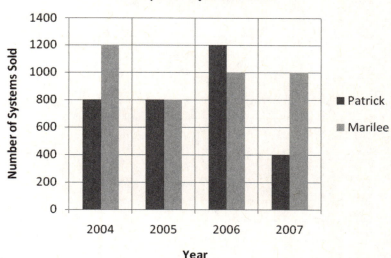

1. In 2005, how many computer systems did sales associate Marilee sell?

   1. _____

2. In 2007, how many computer systems did both sales associates sell combined?

   2. _____

3. Which year was the most successful for the company (based on total systems sold)?

   3. _____

4. In 2004, what percentage of the computer systems did Patrick sell?

   4. _____

5. What was the difference between 2004 and 2007 for total sales?

   5. _____

6. Over the four-year period, how many systems did Patrick sell?

   6. _____

7.   Construct a comparison line graph that compares the number of births during the first six months of the year at two local hospitals.

	Jan.	Feb.	Mar.	Apr.	May	Jun.
Mercy Medical Center	140	120	200	100	150	150
University Memorial Hospital	90	120	100	175	200	125

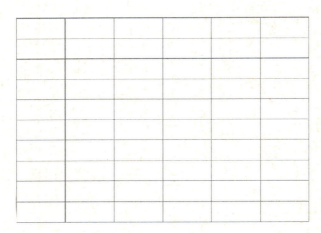

Kelly earns $3000 per month as an administrative assistant. The circle graph divides this monthly salary into basic expense categories. Use this graph to answer the following questions.

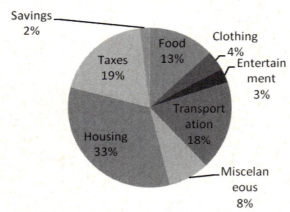

8.   How much does Kelly spend each month on housing?   8. _____

9.   How much more does Kelly spend on taxes compared to clothing?   9. _____

10. How many months will it take Kelly to save $900?   10. _____

Name _____   Date _____

Practice Set 9.2
Mean, Median, and Mode

Find the mean. Round to the nearest tenth when necessary.

1. Chris received the following grades on his tests during the   1. _____
   semester: 73, 82, 88, and 100.

2. Ray, a goalie for the college hockey team, recorded the following   2. _____
   number of saves for the last five games: 31, 37, 25, 16, and 41.

3. The annual salaries of the math department of the local   3. _____
   community college are $64,000, $57,000, $55,000, $67,000, and
   $63,000.

4. Alicia's cell phone bills for the last six months were as follows: 4. _____
   $58.50, $59.70, $39.50, $104.50, $42.50, and $73.90.

Find the median value.

5. The number of cars that turned left at a particular stop sign each   5. _____
   day this week were 118, 198, 175, 98, 155, 145, and 161.

6. The annual salaries of the employees of a small publishing   6. _____
   company are $55,000, $38,000, $36,000, $25,000, $18,000, and
   $27,000.

7. A student's grades for the last six math tests were 85, 82, 80, 90, 7. _____
   79, and 81.

8. The ages of ten people who enter a movie theater are 35, 43, 18,   8. _____
   19, 28, 5, 60, 58, 39, and 33.

Find the mode.

9. Over the last seven days, Holly's Plastic Surgery Center received   9. _____
    the following numbers of inquiries: 45, 55, 65, 55, 25, 60, and
    55.

10. The last eight appliances sold at Ray's Discount Appliances cost   10. _____
    $500, $650, $350, $750, $300, $650, $250, and $350.

11. The last six cars that sold at the used car lot cost $18,500,   11. _____
    $9,500, $12,000, $15,000, $12,000, and $18,600.

12. The teams playing in today's basketball games scored these   12. _____
    numbers of points: 88, 79, 105, 98, 115, 103, 99, 87, 95, 90, 101,
    78, 88, and 93.

13. The points scored by the top two players on the local basketball   13. _____
    team were as follows:

	Game 1	Game 2	Game 3	Game 4	Game 5
Allison	15	18	10	18	14
Mikaela	13	18	18	11	15

    Find the mean, median, and mode for each basketball player.

14. A hot dog stand sold 280, 362, 355, 345, 308, 409, and 307 hot   14. _____
    dogs over a 7-day period. Find the mean, median, and mode.

15. A gas station listed the following gas prices over a 2-week   15. _____
    period: $3.55, $3.28, $3.09, $3.45, $3.49, $3.70, $3.09, $3.45,
    $3.18, $3.60, $3.27, $3.49, $3.25, $3.57. Find the mean, median,
    and mode.

Name _____   Date _____

Practice Set 9.3
The Rectangular Coordinate System

Plot and label the ordered pairs on the rectangular coordinate plane.

1.   (2, 3)

2.   (−3, 5)

3.   (0, 0)

4.   (−2, −3)

5.   (5, −3)

6.   (2, 0)

7.   (−2, 0)

8.   (−1, −1)

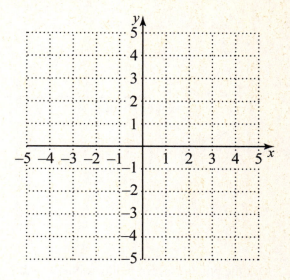

Plot and label the ordered pairs on the rectangular coordinate plane.

9.   (−2.5, 4.5)

10. $\left(3\frac{1}{2}, 1\right)$

11. $\left(4\frac{3}{4}, 0\right)$

12. $\left(0, -2\frac{1}{2}\right)$

13.   (0.5, −3)

14. $\left(-2\frac{1}{2}, -3\frac{1}{2}\right)$

15. $\left(\frac{3}{2}, \frac{10}{3}\right)$

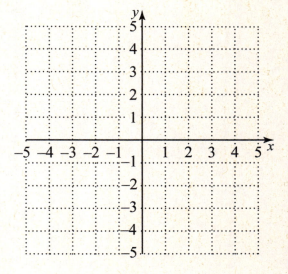

16. Plot the ordered pair that represents 3 miles west and 2 miles south.

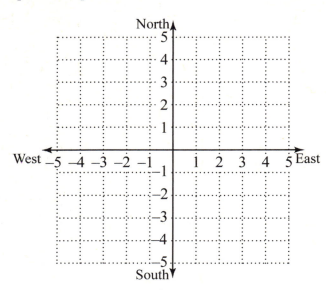

17. Plot the points corresponding to the set of ordered pairs and then draw a line connecting the coordinate points.   (3, –2), (3, 0), (3, –1), (3, 3)

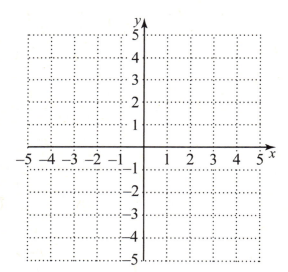

Name _____   Date _____

Practice Set 9.4
Linear Equations in Two Variables

Use a chart to organize your work for Exercises 1–3. Fill in the ordered pairs so that they are solutions to the equations.

1. $x + 2y = 14$ 1.

x	y
2	
	0
6	

2. $x + y = 7$ 2.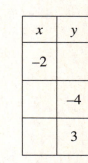

x	y
–2	
	–4
	3

3. $y = x + 3$ 3.

x	y
–5	
	12
0	

Find three ordered pairs that are solutions to the equation.

4. $x - y = -2$ 4. _____

5. $y = 2x - 3$ 5. _____

6. $y = x + 3$ 6. _____

Name _____   Date _____

Plot three ordered-pair solutions to the given equation and then draw a line through the three points.

7. $y = 3x + 2$ 8. $y = -2x + 3$

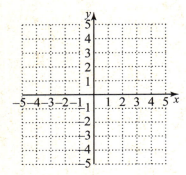

9. $y = 4x + 5$ 10. $y = -x - 1$

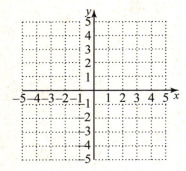
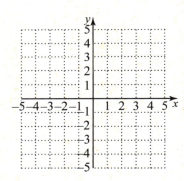

11. $y = -3$ 12. $x = 4$

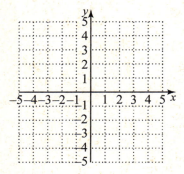
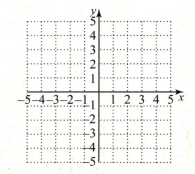

Name _____   Date _____

Practice Set 10.1
Using Unit Fractions with U.S. and Metric Units

Convert.

1.   3 feet = ___ inches   1. _____

2.   ___ pints = 2 quarts   2. _____

3.   ___ cups = 2 pints   3. _____

4.   5280 feet = ___ miles   4. _____

5.   3 days = ___ hours   5. _____

6.   12 tons = ___ pounds   6. _____

7.   88 ounces = ___ pounds   7. _____

8.   ___ gallons = 37 quarts   8. _____

9.   1410 minutes = ___ hours   9. _____

10. 98 cm = ___ mm   10. _____

11. Andrew walked 3.2 kilometers from his house to his friend's   11. _____
     house. How many meters is that?

12. A manufacturer packages ball bearings in crates, each of which   12. _____
     can hold 15,000 grams of ball bearings. How much does each
     crate hold in kilograms?

Fill in the blanks with the correct values.

13. 0.01 L = ___ kL = ____ mL   13. _____

14. 4500 cm = ___ km = ___ m   14. _____

15. 0.09 m = ___ cm = ___ km   15. _____

Practice Set 10.2
Converting Between the U.S. and Metric Systems

Perform each conversion. Round your answer to the nearest hundredth when necessary.

1.   8 ft to m   1. _____

2.   18 m to yd   2. _____

3.   20 km to mi   3. _____

4.   4 in. to cm   4. _____

5.   3.2 mi to km   5. _____

6.   6.7 yd to m   6. _____

7.   0.9 m to ft   7. _____

8.   890 cm to in.   8. _____

Name _____   Date _____

9.   14.5 gal to L   9. _____

10. 3 qt to L   10. _____

11. 5 L to gal   11. _____

Perform each conversion. Round your answer to the nearest tenth when necessary.

12. 45 km/hr to mi/hr   12. _____

13. 30 mi/hr to km/hr   13. _____

Perform each conversion.

14. 35°C to Fahrenheit   14. _____

15. 77°F to Celsius   15. _____

Name _____   Date _____

Practice Set 10.3
Angles

Refer to Figure 1 for Exercises 1–8. Given $p \parallel q$ and $\angle a = 68°$, find the measures of the following angles.

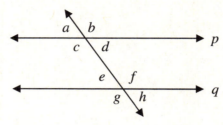

Figure 1

1. $\angle b$ 1. _____

2. $\angle c$ 2. _____

3. $\angle d$ 3. _____

4. $\angle e$ 4. _____

5. $\angle f$ 5. _____

6. $\angle g$ 6. _____

7. $\angle h$ 7. _____

8. $\angle a + \angle b$ 8. _____

Name _____   Date _____

Refer to Figure 2 for Exercises 9–15.

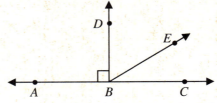

Figure 2

9. $\angle ABC =$ ___°   9. _____

10. $\angle ABD =$ ___°   10. _____

11. $\angle DBC =$ ___°   11. _____

12. $\angle DBE + \angle EBC =$ ___°   12. _____

13. $\angle ABE + \angle EBC =$ ___°   13. _____

14. If $\angle DBE = 50°$, then $\angle EBC =$ ___°.   14. _____

15. If $\angle ABE = 143°$, then $\angle EBC =$ ___°.   15. _____

114

Name _____   Date _____

Practice Set 10.4
Square Roots and the Pythagorean Theorem

Find the square root.

1. $\sqrt{36}$ 1. _____

2. $\sqrt{9}$ 2. _____

3. $\sqrt{16}$ 3. _____

4. $\sqrt{100}$ 4. _____

5. $\sqrt{\dfrac{25}{64}}$ 5. _____

6. $\sqrt{\dfrac{144}{121}}$ 6. _____

Simplify the expressions.

7. $\sqrt{49} - \sqrt{4}$ 7. _____

8. $\sqrt{625} - \sqrt{196}$ 8. _____

9. $\sqrt{81} + \sqrt{169}$ 9. _____

Name _____   Date _____

Solve each applied problem. Round your answer to the nearest tenth.

10. A square patio has an area of 289 square meters. Find the   10. _____
distance between two opposite corners of the patio.

11. Sheila rode her bike 5 miles south and then 2 miles east. How far   11. _____
is she from her starting point?

12. A 22-foot ladder is leaning against a house. The base of the   12. _____
ladder is 4 feet from the foundation of the house. How high up
the building is the ladder resting?

Find the unknown side of each right triangle using the information given. Round your
answer to the nearest thousandth.

13.   13. _____

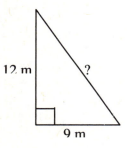

12 m   ?

9 m

14.   14. _____

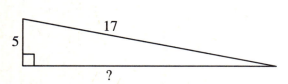

17

5

?

15. One leg of a triangle measures 26 feet, and the hypotenuse   15. _____
measures 38 feet.

116

Name _____   Date _____

Practice Set 10.5
The Circle

Find the circumference of each circle. Use $\pi \approx 3.14$. Round your answer to the nearest tenth.

1. diameter = 14 ft   1. _____

2. radius = 6 in.   2. _____

3. radius = 8.5 in.   3. _____

4. diameter = 140 mi   4. _____

A wheel makes 1 revolution. Determine how far the bicycle travels in inches. Use $\pi \approx 3.14$. Round your answer to the nearest tenth.

5. The diameter of the wheel is 25 in.   5. _____

6. The diameter of the wheel is 4.5 in.   6. _____

A wheel makes 5 revolutions. Determine how far the bicycle travels in inches. Use $\pi \approx 3.14$. Round your answer to the nearest tenth.

7. The diameter of the wheel is 18 in.   7. _____

8. The diameter of the wheel is 42 in.   8. _____

Name _____   Date _____

Find the area of each circle. Use $\pi \approx 3.14$. Round your answer to the nearest tenth.

9. radius = 4 ft   9. _____

10. radius = 16 m   10. _____

11. diameter = 64 cm   11. _____

12. diameter = 3.5 in.   12. _____

Use $\pi \approx 3.14$. Round your answer to the nearest hundredth.

13. A circular, decorative window in a church has a diameter of   13. _____
     72 inches. Due to age, the window needs to have the insulating
     strip that surrounds the window replaced. How many *feet* of
     insulating strip are needed?

14. Dan and Elizabeth would like to construct a concrete circular   14. _____
     patio that measures 15 feet in diameter. They research
     contractors and find that the most economical contractor would
     charge $23 per square yard to pour the necessary concrete. How
     much will this patio cost to construct?

15. A radio station sends out radio waves in all directions from a   15. _____
     tower at the center of the circle of broadcast range. Determine
     how large an area is reached if the diameter of the circle is 140
     miles.

Name _____   Date _____

Practice Set 10.6
Volume

Find each volume. Use $\pi \approx 3.14$. Round each answer to the nearest tenth when necessary.

1. A cylinder with radius 14 in. and height 25 in.   1. _____

2. A sphere with a diameter of 6 in.   2. _____

3. A cone with a height of 30 ft and a radius of 16 ft   3. _____

4. A pyramid with a height of 3.5 ft and a square base of 1.5 ft on   4. _____
    a side

5. A sphere with diameter 12 in.   5. _____

6. A can of soup with a height of 12 centimeters and a radius of 3   6. _____
    centimeters

7. A basketball with a diameter of 20 centimeters   7. _____

8. An ice cream cone, 8 centimeters in diameter and 13 centimeters   8. _____
    in depth, is filled with ice cream level to the top of the cone. How
    much ice cream, in cubic centimeters, is in this cone?

9. A model of a famous pyramid is 8 inches tall. This model has a rectangular base that measures 9 inches by 10 inches. Find the volume of this model.

9. _____

10. Determine the amount of air required to fill a rubber ball with a radius of 7.5 centimeters.

10. _____

11. A collar of Styrofoam is made to insulate a pipe. The pipe has a radius of 6 inches, and the radius of the pipe with the collar is 8 inches. The pipe has a height of 28 inches. Find the volume of the collar.

11. _____

12. An ice cream cone has a height of 7 inches and a radius of 3 inches. It is topped with vanilla ice cream in the shape of a hemisphere. If the cone is completely filled with ice cream, determine the total volume of the ice cream.

12. _____

13. A house is built with a pyramid-shaped roof. The roof has a rectangular base of 8 meters by 6 meters, and a height of 5 meters. The house itself is shaped like a rectangular prism, and has a height of 9 meters. Find the volume of the house.

13. _____

14. A grain silo has a cylindrical body topped by a hemisphere. The cylindrical body has a height of 45 feet and a radius of 8 feet. Find the volume of the grain silo.

14. _____

15. A box is 12 inches long, 8 inches wide and 3 inches tall. It is topped by a cylinder with a diameter of 6 inches and a height of 15 inches. Find the combined volume of the box and cylinder.

15. _____

Name _____   Date _____

Practice Set 10.7
Similar Geometric Figures

For each pair of similar triangles, find the missing side *n*. Round your answer to the nearest tenth when necessary.

1.

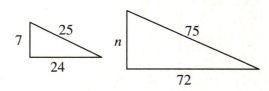

1. _____

2.

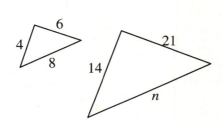

2. _____

3.

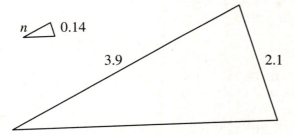

3. _____

4.

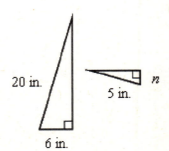

4. _____

Name _____   Date _____

Each pair of figures is similar. Find the missing side. Round to the nearest tenth when necessary.

5.

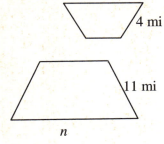

5. _____

6.

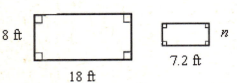

6. _____

7. Tyler and Cori are planning to enlarge their sunroom. The old   7. _____
    sunroom is 8 feet wide by 18 feet in length. The proposed
    sunroom is similar in shape and is 27 feet in length. What is the
    width of the proposed sunroom?

8. A tree casts a shadow of 6 feet. At the same time, a building that   8. _____
    is 50 feet tall has a shadow that is 30 feet. How tall is the tree?

9. Two triangles are similar. The larger triangle has sides of 18 cm,   9. _____
    21 cm, and 28 cm. The 28-cm side on the larger triangle
    corresponds to a side of 21 cm on the smaller triangle. To the
    nearest hundredth, what is the perimeter of the smaller triangle?

10. Keenan is rock climbing in Colorado. He is 6 ft tall and his   10. _____
    shadow measures 9 ft long. The rock he wants to climb casts a
    shadow of 540 ft. How tall is the rock he is about to climb?

11. Paulina took a photo of a panoramic setting during her vacation.   11. _____
    She takes the original 4 in. by 6 in. photo to a photographic
    studio where it will be blown up to poster size, which is 3.5 ft
    tall. To the nearest tenth of a foot, what is the smaller dimension
    (width) of the poster?

122

1.1

1. five thousand three hundred twenty-one; 5000 + 300 + 20 + 1
3. four hundred thirteen thousand, two hundred four; 400,000 + 10,000 + 3000 + 200 + 4
5. 8 hundred-dollar bills and 7 one-dollar bills
7. <
9. 1,000,000 > 0
11. 2012 > 1906
13. 460
15. 2900

1.2

1. $y + 3$
3. $(5 + 6) + x = 11 + x$
5. 13
7. 100
9. 9149
11. 1622
13. 18,614
15. 52 ft

1.3

1. $9 - 8 = 1$
3. $123 - 36 = 87$
5. 2
7. 34
9. 3691
11. 41,102
13. $630
15. 685

1.4

1. associative property
3. $7 \cdot 8$
5. $4 \cdot 2$
7. $x \cdot 3$ (Note: Choice of variable may vary.)
9. 14,000
11. 7,528,014
13. 221
15. 144,000

1.5

1. 2
3. undefined
5. 137 R2
7. 98 R5
9. $36 \div 9$
11. $52 \div x$ (Note: Choice of variable may vary.)
13. 2
15. 12

1.6

1. 3^5
3. 10^4
5. 9^2
7. y^6
9. 512
11. 136
13. 25
15. 32

1.7

1. $3x + 6$
3. $8(x - 1)$
5. 22
7. 10
9. 11
11. $5a + 5$
13. $8a - 8 - 16b$
15. $4x + 5y + 8$

1.8

1. $17a$
3. $13xy$
5. $18ab + 11$
7. The product of four and a number is twenty-four. (Note: Answers may vary.)
9. 6
11. 3
13. $3 + x = 12$
15. $x + (5 + 8) = 16$

1.9

1. $200; $300; $100; $500
3. $400
5. 181
7. $12.53
9. 6
11. $595
13. 120,750

Practice Sets Answers Chapter 2

2.1

1. (number line: points at −5, −2, 0, 3, 4)

3. (number line: points at −3, −1, 1, 2, 4)

5. −100; 100
7. >
9. >
11. −5
13. 42
15. −4; The opposites are 4 and −5, and 4 is greater. (Note: Explanations may vary.)

2.2

1. −8
3. −5
5. 0
7. −1
9. −7
11. 70
13. −2°
15. 175 feet

2.3

1. −5
3. 23
5. −16
7. −8
9. −21
11. −7
13. 98°
15. 5:00 P.M.

2.4

1. $(-10) + (-10) = -20$
3. $(-1) + (-1) + (-1) + (-1) + (-1) + (-1) = -6$
5. 45
7. −125
9. −1
11. −10
13. −77
15. −216

2.5

1. 28
3. −27
5. 35
7. 7
9. 3
11. −14
13. 10
15. $16(14) + 13(12) + 11^2 = 501$

2.6

1. $-13a$
3. $20 + 3t$
5. $-3a - 6ab + 5b + 109$
7. 50
9. −96
11. $3t + 15$
13. $-7a - 28$
15. $28x + 69$

3.1

1. -5
3. -124
5. 13
7. 28
9. 15
11. 38
13. 12
15. $98°$

3.2

1. $T = 2J$
3. $B = 10{,}000Y$
5. 18
7. -13
9. $\dfrac{1}{3}$
11. -9
13. 23 feet
15. 20 minutes

3.3

1. $P = 16$ ft; $A = 15$ ft^2
3. 285 in.^{2}
5. 18 m
7. 80 ft^3
9. 234
11. $\$1160$
13. 16 ft^3
15. 7 ft

3.4

1. 4^5
3. 10^{18}
5. y^{20}
7. $24a^7$
9. $3x^3 + 12x$
11. $-18a^6 + 42a^5$
13. $-4x^4 - x^3y + 20x^3 + y$
15. $198y^2 - 220y$

4.1

1. 2, 3, 5, 7, 11, 13, 17, 19, 23, 29
3. 3 and 5
5. none
7. neither
9. prime
11. $2^2 \cdot 11$
13. $3 \cdot 5^2$
15. $2^2 \cdot 3 \cdot 5^2$

4.2

1. $\dfrac{3}{8}$
3. $\dfrac{13}{25}$
5. $2\dfrac{2}{3}$
7. $10\dfrac{3}{8}$
9. $6\dfrac{6}{11}$
11. $\dfrac{19}{5}$
13. $\dfrac{58}{7}$
15. $\dfrac{119}{5}$

4.3

1. 6
3. 35
5. $6x$
7. $\dfrac{1}{5}$
9. $-\dfrac{3}{5}$
11. $\dfrac{1}{2}$
13. $\dfrac{3}{4}$
15. $\dfrac{19}{40}$

4.4

1. 3^6
3. $\dfrac{1}{x^7}$
5. $\dfrac{2b^2}{3a^2}$
7. $\dfrac{5y^8}{7z^4}$
9. b^{20}
11. $2^3 x^9 y^6 z^3$
13. $\dfrac{2^4}{3^4}$
15. $2^3 3^2 y^{12} z^4$

4.5

1. $\dfrac{3}{1}$

3. $\dfrac{11}{2}$

5. $\dfrac{2}{3}$

7. $\dfrac{5}{7}$

9. 10

11. $8.50 per shirt

13. Cathleen

15. six-ounce

4.6

1. =

3. =

5. 9

7. 10

9. 169

11. $44

13. 70 minutes

15. $4\dfrac{1}{2}$ kilometers

5.1

1. $\dfrac{2}{45}$

3. $\dfrac{33}{100}$

5. $\dfrac{13}{14}$

7. $\dfrac{7x^2}{2}$

9. $\dfrac{18}{25}$

11. -2

13. $\dfrac{7x^3}{6}$

15. $12x^7$

5.2

1. 14

3. 36

5. 180

7. 42

9. $32x^2$

11. $150x^4$

13. 11:55 A.M.

15. 12:06 A.M.

5.3

1. $\dfrac{7}{11}$

3. $-\dfrac{1}{2}$

5. 4

7. $\dfrac{7}{10}$

9. $-\dfrac{11}{24}$

11. $\dfrac{29}{60x}$

13. $\dfrac{2a+5b}{ab}$

15. $\dfrac{23}{40}$

5.4

1. $15\dfrac{2}{3}$

3. $4\dfrac{1}{7}$

5. $17\dfrac{2}{5}$

7. $11\dfrac{8}{11}$

9. $7\dfrac{1}{2}$

11. $12\dfrac{1}{40}$

13. $-11\dfrac{1}{3}$

15. $1\dfrac{12}{23}$

5.5

1. $84\frac{1}{3}$

3. $\frac{1}{125}$

5. $\frac{8}{9}$

7. $\frac{7}{100}$

9. 20

11. $1\frac{1}{4}$

13. $\frac{25}{28}$

15. $\frac{1}{2}$

5.6

1. 50 mph

3. $5\frac{1}{2}$ yards

5. 9 lb of meat, $16\frac{1}{5}$ lb of potato salad,

   $13\frac{1}{2}$ lb of fruit

7. $17\frac{5}{11}$

9. 5

11. $106\frac{2}{3}$

5.7

1. 36
3. 84
5. 96
7. 275
9. −250
11. 33
13. −112
15. 35

6.1
1. $+3x^2, +4xy, +5y^2$
3. $3x - 6$
5. $7y^2 - 8y - 1$
7. $-6x - 10$
9. $12y - 1$
11. $-3a + 28$
13. $2y^3 - y^2 - 6$
15. $5x^5 - 3x^4 + 3x^3 - 2x^2 - x + 7$

6.2
1. $15x^2 + 10x - 20$
3. $-20x^5 - 8x^4 + 20x^2$
5. $2x^3 + 5x^2 - x - 1$
7. $a^2 + 5a + 6$
9. $y^2 - 17y + 60$
11. $6x^2 - 8x - 8$
13. $-6x^3 + 7x^2 - 5x - 15$
15. $-2x^3 - 5x^2 - x - 12$

6.3
1. w = width, $3w$ = length
3. A = Anne Marie's monthly salary,
   $A + \$125$ = Raymond's monthly salary
5. A = Andy's height, $A + 5$ = Kari's
   height, $A + 3$ = Sharon's height
7. R = Ray's winnings, $2R$ = Chris's
   winnings, $R - \$35$ = Jay's winnings
9. A = fish in Pond A, $A - 40$ = fish in
   Pond B, $2A + 15$ = fish in Pond C
11. a. H = Sheldon's hockey cards,
       $H + 350$ = Brandon's hockey cards,
       $H - 180$ = Trevor's hockey cards
    b. $(H + 350) + (H - 180) - H$
    c. $H + 170$

6.4
1. ab^2
3. 1
5. $7(x + 1)$
7. $3(x - 3)$
9. $5(a - 9)$
11. $2(3a + 4b + 7)$
13. $2y(4y - 3)$
15. not factorable; GCF is 1.

131

7.1

1. −7
3. −38
5. 18
7. −5
9. 11
11. 16
13. $\dfrac{5}{6}$
15. −33

7.4

1. 18
3. $\dfrac{20}{3}$
5. −15
7. $\dfrac{1}{4}$
9. $\dfrac{7}{24}$
11. 6
13. $-\dfrac{12}{5}$ or $-2\dfrac{2}{5}$
15. $-\dfrac{10}{7}$ or $-1\dfrac{3}{7}$

7.2

1. 1
3. 0
5. −2
7. $\dfrac{15}{2}$
9. 2
11. $\dfrac{13}{5}$
13. $\dfrac{16}{5}$
15. 5

7.5

1. 13 m
3. 14 ft
5. a. 5 ft
   b. 9 ft
7. 7 ft
9. John $200; Kevin $250

7.3

1. 3
3. −6
5. −1
7. −2
9. −36
11. −19
13. 6
15. −4

8.1

1. three and forty-eight hundredths

3. twenty-two and $\dfrac{43}{100}$

5. $\dfrac{4}{10}$

7. $5\dfrac{327}{1000}$

9. 0.3

11. 19.243

13. >

15. 3.142

8.2

1. 10.8
3. 10.04
5. 28.374
7. 8.6
9. 73.281
11. −6.4
13. $13.9x - 2.1y$
15. 12.26

8.3

1. 0.0018
3. 0.0012
5. −45.69
7. 2368.1
9. −0.014
11. 3.25
13. 2.43
15. $11.2\overline{6}$

8.4

1. 6.1
3. 9.8
5. 6
7. 0
9. −3.8
11. −8.125
13. 17
15. Yes, he had $13.05 left over.

8.5

1. 6
3. 30
5. 120
7. 30
9. 240
11. 84,000
13. $6
15. $14,000

8.6

1. 47%
3. 123%
5. 0.37
7. 0.00017
9. 38.5%
11. 7%
13. 533.33%
15. $\dfrac{1}{800}$

8.7

1. 60
3. 387.2
5. 128.65
7. 2300
9. 86
11. 15%
13. 6%
15. 80%

8.8

1. 24, 37, 9
3. p, 350, 35
5. 88, 198, a
7. 63
9. 300
11. 800
13. 40%
15. 10%

8.9

1. $8000
3. $8360
5. $29.20
7. a. $144
    b. $2544
9. 90

Practice Sets Answers Chapter 9

9.1

1. 800
3. 2003
5. 600 systems
7.

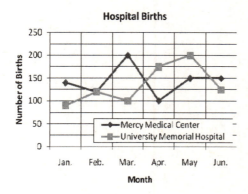

9. $450

9.2

1. 85.8
3. $61,200
5. 155
7. 81.5
9. 55
11. $12,000
13. Allison: 15, 15, 18; Mikaela: 15, 15, 18
15. Mean: $3.39, Median: $3.45, Mode: $3.09, $3.45, $3.49

9.3

1.-7.

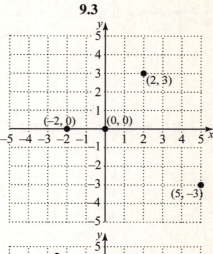

9.-15.

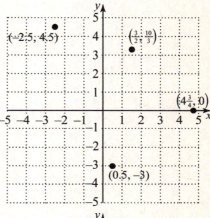

17.

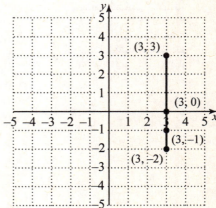

9.4

1. 6; 14; 4

3. −2; 9; 3

5. Answers will vary. (0, −3), (1, −1), (3, 3)

7.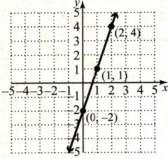

9.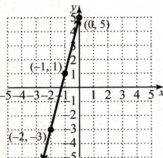

11.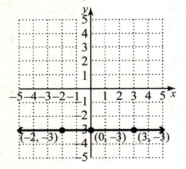

136

10.1

1. 36
3. 4
5. 72
7. 5.5
9. 23.5
11. 3200
13. 0.00001, 10
15. 9, 0.00009

10.2

1. 2.44 m
3. 12.43 mi
5. 5.15 km
7. 2.95 ft
9. 54.89 L
11. 1.32 gal
13. 48.3 km/h
15. 25°C

10.3

1. 112°
3. 68°
5. 112°
7. 68°
9. 180
11. 90
13. 180
15. 37

10.4

1. 6
3. 4
5. $\frac{5}{8}$
7. 5
9. 22
11. 5.4 mi
13. 15 m
15. 27.713 ft

10.5

1. 44.0 ft
3. 53.4 in.
5. 78.5 in.
7. 282.6 in.
9. 50.2 ft^2
11. 3215.4 cm^2
13. 18.84 ft
15. 15,386 mi^2

10.6

1. 15,386 in.^{3}
3. 8038.4 ft^3
5. 904.3 in.^{3}
7. 4186.7 cm^3
9. 240 in.^{3}
11. 2461.8 in.^{3}
13. 512 m^3
15. 711.9 in.^{3}

10.7

1. 21
3. 0.3
5. 24.8 mi
7. 12 ft
9. 50.25 cm
11. 2.3 ft